Science, Revolution and Discontinuity

Science, Revolution and Discontinuity

JOHN KRIGE
*Lecturer in History and Social Studies of Science,
University of Sussex*

THE HARVESTER PRESS · SUSSEX
HUMANITIES PRESS · NEW JERSEY

First published in Great Britain in 1980 by
THE HARVESTER PRESS LIMITED
Publishers: John Spiers and Margaret A. Boden
16 Ship Street, Brighton, Sussex
and in the USA by
HUMANITIES PRESS INC.,
Atlantic Highlands, New Jersey 07716

© 1980 John Krige

British Library Cataloguing in Publication Data
Krige, John
  Science, revolution and discontinuity –
  (Harvester studies in philosophy; no.10)
  1. Science – Philosophy
  2 – Science – History
  I. Title
  509   Q175

ISBN 0-85527-625-8

Humanities Press
ISBN 0-391-02094-3

Printed in Great Britain by
Redwood Burn Ltd., Trowbridge and Esher

All rights reserved

For David and Peter

CONTENTS

	Page
Introduction	9
1 Revolution and discontinuity	15
2 The poverty of Popperism	45
3 Lakatos' quest for continuity	70
4 From discontinuity to Dadaism	106
5 Structures and impossibilities	157
6 Impetus, inertia and discontinuity	175
7 Concluding remarks	214
Bibliography	221
Index	229

INTRODUCTION

It is all too easily forgotten that revolutionary changes are a major concern of some of the most important contemporary philosophers and historians of science. The copious literature surrounding the work of Popper, Lakatos, Feyerabend and Kuhn frequently fails to recognize that, as Kuhn puts it, both he and his Popperian critics 'share the conviction that the central episodes in scientific advance — those which make the game worth playing and the play worth studying — are revolutions'.[1]

To characterize one's project in this way is, of course, to assume that revolutions have occurred in science. This assumption has not gone unchallenged. Among philosophers of science Toulmin, for example, has consistently queried it. Historians, most notably Duhem, are also sceptical. Their reluctance is informed by the continuities which they have detected in the historical record. In particular, the apparently unbroken line of descent leading from at least fourteenth century impetus theory through Galilean physics to the 'classical' principle of inertia has led them, and others, to doubt that there was a revolution in science in the seventeenth century, as is commonly believed.

Revolution and discontinuity, then, go hand in hand. And insofar as one is concerned with the development of scientific *thought*, it is typically the nature and extent of *conceptual* transformation that serve as an index of revolutionary change. This point is important. For as Laudan, for one, has noted, presently dominant empiricist epistemologies tend to concentrate on the role which the solution of empirical problems has played in scientific advance. This has often been done at the expense of analyzing the conceptual changes which are also an essential part of the growth of knowledge. As a matter of fact, the solution of certain particularly intractable empirical problems may itself require conceptual innovation of a fairly drastic kind.

The question of just how deep or far reaching conceptual change in science has been, or need be, is the question of whether there have been, or need be, discontinuous transitions in the development of scientific thought. Historically speaking, has the advance of knowledge involved, perhaps even demanded, the production of conceptual frameworks which are *so* different from their predecessors, that there is no effective overlap between them ? Or, in language which is rather more familiar to contemporary philosophers of science, have there been transitions between incommensurable systems of thought which, for that reason, are to be regarded as discontinuous transitions ?

Foucault is one historian who believes that transformations of this type have indeed occurred. For example, he invites us to consider

> medicine at the end of the eighteenth century: read twenty medical works, it doesn't matter which, of the years 1770 to 1780, then twenty others from the years 1820 to 1830, and I would say, quite at random, that in forty or fifty years everything had changed; what one talked about, the way one talked about it, not just the remedies of course, not just the maladies and their classifications, but the outlook itself . . . [This change was] a matter of a collective and complex transformation of medical understanding in its practice and its rules . . . it represents the application of an entirely new *grille*, with its choices and exclusions; a new play with its own rules, decisions and limitations, with its own inner logic, its parameters and its blind alleys, all of which lead to the modification of the point of origin.[2]

In similar vein Koyré insists that a decisive mutation occurred in physics in the seventeenth century, and that classical physics was not in fact continuous with medieval physics. It is this, he says

> which explains why the discovery of things which seem to us nowadays childishly simple required such prolonged efforts — and not always crowned with success — by the greatest of geniuses, by Galileo and Descartes. This is because it was not a matter of battling against theories which were simply inadequate or erroneous, but of changing the very intellectual framework itself, of overthrowing an intellectual attitude, one which was when all is said and done a perfectly natural one, and substituting for it another, one which was not natural at all.[3]

If we accept Foucault's and Koyré's reading of the historical record we are forced to reconsider the relationship between revolution and discontinuity. It is one thing to detect discontinuities and another to assert that the process or mechanism whereby they took place was a revolutionary one. 'The revolutionary process', as Kuhn has pointed out, is one 'by which an older theory is rejected and replaced by an incompatible new one'.[4] But it is more than this. For revolutions are conventionally regarded as more or less sudden, catastrophic and rationally inexplicable outbursts of energy which precipitate the discontinuous changes alluded to by Kuhn. And since the transformations in understanding of which Foucault and Koyré speak demanded a great deal of time and effort, there is some question as to whether or not they are to be regarded as revolutions even though they were discontinuous transitions.

This then is the difficulty with which we are faced. Historical studies suggest that at times one conceptual framework has completely replaced a predecessor which was incommensurable with it. The transition in question was thus discontinuous. Furthermore the comprehensive nature of the change, involving rejection and replacement of the old by the new, suggests that the transformation in question was revolutionary. On the other hand the transition is not sudden; it is necessary to consider two universes of discourse separated by a considerable period of time to detect historical discontinuities at all. In the short term a measure of conceptual continuity is as evident as is a certain amount of conceptual discontinuity. In conjunction with the view of revolutions which takes them to be sudden, cataclysmic events, it appears then that insofar as a discontinuous transition occurred at all, it was not a revolutionary one. Correlatively, it seems that even fairly drastic changes are essentially gradual, piecemeal and evolutionary.

It is my contention that the dilemma just identified is a false one, and is precipitated by an inadequate conception of revolution. It is that conception which takes revolutions to be explosive, rationally unintelligible events which destroy an established order *in one fell swoop*, replacing it with a new and quite different system or pattern of relationships. My aim in this book is to develop a conception of the revolutionary process which does not suffer from this disadvantage, and which assumes as a working hypothesis that such transformations are rationally intelligible.

Revolutionary changes are still regarded as being discontinuous, and as involving a double aspect; the *rejection* of one framework and its *replacement* by another. But it is stressed that these two aspects of a revolutionary transition are, so to speak, out of phase with each other. Rejection and replacement, in other words, do not occur more or less simultaneously. The old order is rejected and the *foundations* of the new are laid fairly rapidly. However, it *takes time* for the latter to *replace* the former. That granted, if the commitment to revolutionary change is to be regarded as rational, a conception of rationality is required which reflects the 'asymmetrical' and protracted nature of the process.

To my mind a cardinal flaw in contemporary theories of rational scientific growth lies in their inability to deal with this central feature of revolutionary change. This emerges from the discussion in Chapters 2 to 4 below, which deal respectively with the views of Popper, of Lakatos and of Feyerabend. At one level these chapters simply describe the work of the three philosophers in question, who are treated primarily as theorists of revolutionary change. As such, this part of the book can be considered as an introduction to their views from a particular vantage point. However, it is also stressed that, after Popper, there is a movement (through Lakatos to Feyerabend) away from the idea that one can ever rationally reject a scientific theory (ultimately) replacing it with a new and better alternative. It follows that, insofar as discontinuous changes have occurred at all, the trend amongst these philosophers of science is to suggest that such changes are not rationally explicable or justifiable.

This trend is traced back to the more or less explicit adoption of a distinction between the context of discovery and the context of justification. Allegiance to this distinction has marked effects on the way in which one approaches the question of theory acceptance and rejection. In particular, the conditions under which, and the processes whereby, new knowledges are produced are located in the context of discovery. This amounts to assuming that they are not amenable to rational explanation. It is only the choice between fully-fledged systems of thought that can be appraised from a rational point of view — an exercise restricted to the so-called context of justification.

It is one of the central claims of this book that as long as philosophers remain trapped within this presuppositional framework they have no hope of rendering revolutionary changes in science rationally intelligible. This is because of the fundamental asymmetry of the revolutionary process to which I alluded above. In a revolutionary situation one does not choose between two fully articulated systems of thought (or modes of social organization). Science, as Feyerabend likes to say, develops unevenly. One rejects a 'fully articulated' system, and lays the foundations for a new system which is initially woefully inadequate when compared with its well entrenched rival.

It is my contention that part of what makes it reasonable for people to cling tenaciously to an embryonic alternative is the conviction that the old order has exhausted all the possibilites for development inherent in it. More specifically, the belief that the established framework not only *has not* but actually *cannot* solve the problems with which it is confronted is a prerequisite for its rational rejection. It is partly that conviction which inspires revolutionary agents with the confidence that they need to produce an alternative which can ultimately replace the orthodoxy of the day.

This conception of the nature of revolutionary change is based primarily on the historical analyses presented in Chapters 5 and 6. These chapters rely heavily on the work of Koyré and of Clavelin. They explore the basis for, and the implications of, Galileo's conviction that the Ptolemaic/ Aristotelian system was, as he put it, 'undoubtedly false', whereas the Copernican system 'may be true'. Taken together, these two assertions illustrate the asymmetrical nature of revolutionary commitment which I am arguing for. Galileo's rejection of Aristotelianism was an essential dimension of his commitment to Copernicanism. Any attempt to explain as rational his espousal of the heliostatic system which overlooks or plays down his determined opposition to established orthodoxy is, to that extent, one-sided and inadequate. In Chapter 5 the grounds for Galileo's rejection of Aristotelianism are investigated. Chapter 6 discusses one important aspect of his defence of Copernicanism — the development of the principle of inertia — and analyzes the difficulties which he encountered.

The body of the book, part philosophical, part historical is framed by an introductory chapter and a conclusion. The fomer expands somewhat discursively on the theme of revolution and discontinuity, with special emphasis on the views of historians of science. Kuhn's conception of revolutionary change is spelt out in some detail. It is contrasted with a quite different conception drawn from a variety of thinkers, including Marx. This alternative, it is suggested, more adequately theorizes the kind of discontinuous transition which Koyré, for example, has described. In the conclusion as much mileage as possible is extracted from the historical material in Chapters 5 and 6. The argument in the book is summarized, and some of its implications for philosophers and historians of science who are concerned with revolutionary transformations are discussed.

This book is based on my doctoral dissertation submitted to the University of Sussex in September 1978. As such it owes much to the unwavering encouragement and to the ceaseless intellectual stimulation of my supervisor, Roy Edgley. Maggie Boden and Jerry Ravetz read the text in draft, and made many valuable suggestions which have improved it. My thanks, too, are due to my wife who supported us financially for several years while I was engaged in the research which led to my doctorate. The thesis was ultimately produced during a period of intense personal upheaval and strain. Without the support of the many men, women and children who sustained me at that time, it would not have emerged in its present form, if at all.

Notes

1 T.S. Kuhn, 'Reflections on my critics', in I. Lakatos and A. Musgrave (eds), *Criticism and the Growth of Knowledge* (Cambridge University Press, Cambridge, 1970), p.241.
2 M. Foucault, 'Human nature: justice versus power', debate with N. Chomsky in F. Elders (ed.), *Reflexive Waters*, (Souvenir Press, London, 1974), p.150.
3 A. Koyré, *Galileo Studies*, J. Mepham (trans), (Harvester Press, Hassocks, 1978), p.3.
4 T.S. Kuhn, 'Logic of discovery or psychology of research ?', in I. Lakatos and A. Musgrave (eds.), *op. cit.* p.2.

1 REVOLUTION AND DISCONTINUITY

IN what sense, if at all, can the achievements of men like Copernicus, Galileo, Descartes and Newton be said to constitute a scientific *revolution*? The hesitancy which informs this question is a symptom of the caution which many historians and philosophers now feel when discussing the birth of modern science. They no longer have the confidence of a Kant who, in his Preface to the second edition of the *Critique of Pure Reason*, enthusiastically proclaimed that a 'new light flashed upon all students of nature' in the seventeenth century. And they are far less convinced than he was that physics underwent a 'rapid intellectual revolution' at the hands of Galileo and his contemporaries. Studies of innovations made in the fourteenth century, particularly at Oxford and at Paris, suggest that some of the most fundamental concepts and methods of classical science were anticipated by the medieval schoolmen. It appears that they were not 'groping in the dark', unaided by the light of reason, as Kant supposed — for example, some of them believed, like Galileo, that nature could be comprehended by mathematical laws. The view that the (so-called) Scientific Revolution involved a clean break with the past is wearing thin. Many would agree with Crombie that 'the most striking result of recent scholarship (is) the essential continuity of the Western scientific tradition from Greek times to the seventeenth century and, therefore, to our own day'.[1]

If Crombie appears to favour a continuist theory of history here, there are hints elsewhere in the same text that suggest that he has not really thought out his position on this matter too carefully. Thus, he also says that during the sixteenth and seventeenth centuries the successful application of mathematics to mechanics 'changed men's whole conception of Nature and brought about the destruction of the whole Aristotelian system of cosmology'.[2] He is not alone in insisting that a profound and far-reaching change occurred at this time. Hall, for example, takes care to distinguish the emergence of classical mechanics from the transition from Newtonian to

Einsteinian physics at the turn of this century, when he says that

> Only the broader extrapolations of nineteenth-century science would now be described blankly as 'wrong', though a larger part of its picture of Nature might be described as 'inadequate' or as 'true within certain limits' . . . . The same could not be said of science before 1500, or even, without restriction, of the science of the seventeenth and eighteenth centuries. Its progress in these earlier times was not by accretion, for it was now and again necessary to jettison encumbering endowments from the past. Such science *was* on occasion simply wrong, both in fact and in interpretation. Its propositions had to be rejected *in toto*, not merely circumscribed, as the result of experiment and creative thinking.[3]

Thus Hall, like Crombie, seems to think that the Scientific Revolution was discontinuous in the sense that one conception of nature was completely replaced by another.

This tension — the tension between the search for continuity and the recognition that in the Scientific Revolution an entire system of thought seems to have been obliterated — reaches breaking point in Butterfield's Introduction to his study of the origin of modern science. Consider the contradictions — as revealed by my emphases — in the following quotation from that Introduction. There is immense benefit to be gained, says Butterfield, from a close study of

> the *so-called 'scientific revolution'*, popularly associated with the sixteenth and seventeenth centuries, but reaching back in an *unmistakably continuous* line to a period much earlier still. Since that *revolution overturned* the authority in science not only of the middle ages but of the ancient world — since it ended not only in the *eclipse* of scholastic philosophy but in the *destruction* of Aristotelian physics — it outshines everything since the rise of Christianity and reduces the Renaissance and Reformation to the rank of mere *episodes*, mere *internal displacements*, within the system of medieval Christendom . . . . It changed the character of man's habitual mental operations even in the conduct of the non-material sciences, while transforming the *whole* diagram of the physical universe and the very texture of human life itself.[4]

Butterfield, as we can see, begins by defending continuity, hesitating to call the Scientific Revolution a revolution, and ends up by undermining his own position completely.

In this quotation from Butterfield two seemingly opposed accounts of the transition from Aristotelian thought to modern science jostle with one another. Both are due primarily to the labours of two French historians and philosophers: Pierre Duhem and Alexandre Koyré. Around the turn of the century Duhem concluded, on the basis of an extensive study of medieval manuscripts, that, as Drake puts it

> it is possible to look upon science as having undergone a continual development and transformation from Greek antiquity to the present. Progress was at an uneven rate, in that view, but not discontinuous; the so-called Scientific Revolution of the seventeenth century belonged more properly to the thirteenth and fourteenth centuries, and its sources there lay still farther back.[5]

In the light of investigations made during the 1930s, Koyré came to a quite different conclusion. 'The apparent continuity in the development of medieval and modern physics', he writes '(a continuity so emphatically stressed by Caverni and Duhem) is an illusion'.[6] Koyré regards the Scientific Revolution as having involved an intellectual 'mutation' — a term borrowed from Bachelard — as profound as any in human thought. Whereas historians of continuity find themselves in the embarrassing position of seeming to play down the achievements of Galileo and of his successors, Koyré emphasizes their originality. As far as he is concerned we can only appreciate the extent of their efforts if we realize that what they had to do

> was not to criticize and to combat certain faulty theories, and to correct or to replace them by better ones. They had to do something quite different. They had to destroy one world and to replace it by another. They had to reshape the framework of our intellect itself, to restate and to reform its concepts, to evolve a new approach to Being, a new concept of knowledge, a new concept of science — and even to replace a pretty natural approach, that of common sense, by another which is not natural at all.[7]

The choice between Duhem's position and Koyré's seems to be quite straightforward. Yet as we have seen eminent historians like Crombie and Butterfield oscillate between the continuist and discontinuist poles without apparently realizing that they are doing so. Some scholars simply adopt a cosy compromise. Drake, for example, suggests that 'Whether we wish to look upon the history of science as continuous or discontinuous throughout the centuries depends more on philosophical prepossessions than it does on the record of ideas and events. Either approach is potentially fruitful of increased understanding, but not at the expense of the other'.[8]

Wiener and Noland, on the other hand, reassure their readers that 'the intellectual lines of continuity between ancient and modern science are there despite the revolutionary changes that have occurred within the history of science'.[9] This quest for continuity is, it would seem, compulsive for some historians. Dijksterhuis concludes his summary of the 'evolution' of mechanics from Greek to medieval times with the remark that his analysis 'greatly satisfies our desire for the combination of continuity and renewal, a desire which as a rule spontaneously directs our historical inquiries'.[10]

My aim in this chapter is to explore, albeit somewhat tentatively, some of the factors which have produced this rather confused state of affairs. Following Koyré, I shall assume (here, and throughout the book) that there *was* a revolution in science in the seventeenth century and that it involved a profound 'transformation of the human understanding', as Foucault puts it. At a later stage in the analysis I shall argue for these claims; in this introductory survey they are simply taken for granted. As we shall see, they amount to the assertion that the conceptual structures of Aristotelian physics and cosmology, on the one hand, and of classical mechanics on the other, are *so* different from one another that there is no effective overlap between them. They are incommensurable systems of thought, one of which was rejected and replaced by the other.

I am aware that most philosophers of science will spontaneously resist my claim that there can be discontinuous transitions between incommensurable conceptual frameworks. Even someone like Chalmers, who has pretensions to radicalism, unhesitatingly dismisses the idea as 'obscurantist nonsense'.[11] However, it is my contention that this resistance reflects, at least in part, the pressures exerted by a particular conception of revolution and a related conception of rationality. For convenience I shall call this the liberal conception of revolution. As we shall see below, if one approaches the historical record armed with this conception of revolution, one has little option but to deny that discontinuous transitions ever occur. Correlatively, it appears either that all historical change is evolutionary, or that revolutions, if they take place at all, are so localized that they do not substantially disturb the tranquility of an essentially continuous historical process.

In what follows I shall contrast the liberal conception of revolution with what I shall call, again for convenience, a Marxist conception of revolution. It is my contention that an adequate appreciation of the nature of discontinuous transitions requires that the revolutionary process be understood in the 'Marxist' sense.

From the outset I should stress that in this chapter I have sometimes chosen to be less than rigorous rather than omit material that is potentially fruitful and illuminating. By doing so, I hope to make the reader aware of the richness and complexity of discontinuous transitions — phenomena that all too easily elude our gaze as long as we remain trapped within the confines of a stultifying conception of revolution and of rationality.

1.1 The 'Liberal' Conception of Revolution

As is only to be expected, revolutions in society and in science have often been thought about in the same way. No doubt Kant's frequent use of the word 'revolution' in the Preface to the second edition of his *Critique*, which was published in 1787, was indicative of the turmoil which surrounded him. Prior to about 1700 the concept of revolution does not appear to have been applied to scientific changes.[12] In fact it was not even commonly used in political discourse until after the Glorious Revolution of 1688. At this time the concept of revolution invoked in this context was usually that borrowed from astronomy and geometry: it meant a cyclical, continuous process, with the connotation of a return to a former state. During the eighteenth century its meaning changed. Increasingly it came to signify a somewhat drastic transformation of the social and political order, possibly accompanied by violence. This new sense of the term was then fed back into accounts of scientific change itself, which was 'now visualized as a series of secular discontinuities of such magnitude as to constitute definite breaks with the past'.[13] It was this, still prevalent, meaning of the term which rose to prominence at about the time when Kant was writing — after 1789, in fact.

More than an allusion to far-reaching change is built into the conception of discontinuity inspired by the French (and American) revolutions. It is also implied that one system is effectively wiped out and replaced by another bearing no resemblance to it in a very short space of time. In 1890 Berthelot wrote of Lavoisier's achievements in chemistry in this vein. He compared the political and scientific events of revolutionary France with one another. The former 'reconstituted the society among us on new foundations, and demarcated an era basically new in the history of humanity'.[14] Simultaneously there was a 'considerable revolution' in the sciences, including chemistry. And in both cases the transformations 'were not effected gradually, through the slow evolution of years and the accumulated work of several generations of thinkers. No! They were on the contrary produced suddenly: fifteen years were enough to accomplish them'.[15] Narrow in time, with effects as deep as they are wide — it is these features which together constitute, for Berthelot, a rupture with the past.

If anything, some contemporary historians of science have tended to see intellectual breaks with the past as being even more sudden than Berthelot imagined. For Butterfield, for example, the initial transition from Aristotelian dynamics to inertial physics was instantaneous. He attributes this to the specific nature of the relationship that holds between the two systems of thought. In Aristotelian physics it was a body's persistence in motion that needed to be explained; rest was the natural state for all non-celestial bodies. In classical physics, on the other hand, it is the *failure* to persist in a state of uniform motion (or rest) which demands an explanation. Thus Butterfield suggests that inertial physics emerged 'somewhat on the policy of picking up the opposite end of the stick'.[16] But to do that one had to imagine geometrical bodies moving freely in abstract Euclidean space. This novel way of thinking about motion could not be effected, says Butterfield, by observing the movement of real bodies more carefully; 'it required a different kind of thinking-cap, a transposition in the mind of the scientist himself'.[17]

Like Kant, then, Butterfield suggests that a new light

dawned on students of nature in the seventeenth century. However, although he isolates a 'moment' of revolutionary rupture, he does not imply that there was simultaneously a complete break with the past. This is an important distinction, to which I shall return later.

But of course it is in Kuhn's theory of history that we find one of the most comprehensive and radical conceptions of discontinuous change. His views are articulated in opposition to a historiographical tradition that identifies science with the constellation of facts, theories and methods that constitute the standard diet of today's scientists. From this perspective, its development 'becomes the piecemeal process by which these items have been added, singly and in combination, to the ever growing stockpile that constitutes scientific technique and knowledge. And history of science becomes the discipline that chronicles both these successive increments and the obstacles that have inhibited their accumulation'.[18] The past is studied for the sake of the present, to which it is subordinated; the present, as Butterfield points out, is thereby ratified if not glorified. An historian who adopts this 'Whig' interpretation of history, as he calls it,

is bound to construe his function as demanding him to be vigilant for likenesses between past and present, instead of being vigilant for unlikenesses; so that he will find it easy to say that he has seen the present in the past, he will imagine that he has discovered a 'root' or an 'anticipation' of the 20th century, when in reality he is in a world of different connotations altogether, and he has merely tumbled upon what could be shown to be a misleading analogy . . . . The total result of this method is to impose a certain form upon the whole historical story, and to produce a scheme of general history which is bound to converge beautifully upon the present — all demonstrating throughout the ages the workings of an obvious principle of progress.[19]

A continuist history of this type sees scientific knowledge as developing by accretion. It fails to recognize the internal coherence of earlier theories and of different belief systems, which seemed just as well founded and reasonable to their protagonists as do our beliefs to us. As such, it overlooks the discontinuous and revolutionary transitions from one interwoven 'world of connotations' to another quite different one which, Kuhn insists, are revealed by a careful scrutiny of the historical record. More specifically, and typically

Even a theory like energy conservation, which today seems a logical superstructure that relates to nature only through independently established theories, did not develop historically without paradigm destruction . . . . Only after the caloric theory had been rejected could energy conservation become part of science. And only after it had been part of science for some time could it come to seem a theory of a logically higher type, one not in conflict with its predecessors. It is hard to see how new theories could arise without these destructive changes in beliefs about nature. Though logical inclusiveness remains a permissible view of the relation between successive scientific theories, it is a historical implausibility.[20]

Kuhn does not deny that the piecemeal, cumulative conception of scientific development has some merit. In fact, he concedes that it provides a more or less satisfactory account of scientific progress during the, often protracted, periods of history which Kuhn labels 'normal science'. During these periods research is conducted under the restrictive umbrella of a 'paradigm'. Kuhn's conception of a paradigm is notoriously vague and ambiguous. Typically it is a 'strong network of commitments — conceptual, theoretical, instrumental and methodological'[21] which maps out in advance permissible directions in which scientific inquiry is to proceed, and the techniques appropriate to that task. Once socialized into a paradigm, the practitioners of normal science do not question its fundamentals. Rather, equipped with a more or less explicit set of rules for conducting research, they aim to solve puzzles which are generated by the paradigm. Puzzles — for what a paradigm does is to instil in those who accept it the confidence that the difficulties that they encounter in their research can be resolved within its framework. Like the student who tackles illustrative examples at the end of the chapter in a science textbook, they believe that their 'problems' can be solved as long as they follow the guidelines laid down by the paradigm. Normal science, then, is 'the generally cumulative process by which the accepted beliefs of a scientific community are fleshed out, articulated, and extended. It is what scientists are trained to do . . . . .'.[22]

The hegemony of normal science is not total, however, nor is its reign interminable. There are always mismatches between paradigm-induced expectations and the actual state of affairs. These anomalies are initially brushed aside. Some are solved with the passage of time; others become naggingly persistent. Under the pressure exerted by the build-up of anomaly, scientists' commitment to their paradigm falters. A state of crisis ensues. Fundamentals are questioned, and a number of competing theories are put forward: we have entered a period of extra-ordinary science (or, in some of Kuhn's formulations, of non-science or metaphysics). The crisis is terminated when one of the competitors is espoused at the expense of its rivals, and a new paradigm is established. With its advent, the terms of scientific debate are changed, often so drastically that only a new generation of practitioners has the mental flexibility required to espouse it.

It is these 'tradition-shattering complements to the tradition-bound activity of normal science'[23] that Kuhn calls scientific revolutions. And he is emphatic that they cannot be accomodated within a theory of history that sees science as developing in a piecemeal, cumulative way, for he says that

> The transition from a paradigm in crisis to a new one from which a new tradition of normal science can emerge is far from a cumulative process, one achieved by an articulation or extension of the old paradigm. Rather it is a reconstruction of the field from new fundamentals, a reconstruction that changes some of the field's most elementary theoretical generalizations as well as many of its paradigm methods and applications.[24]

For Kuhn, then, revolutions are 'destructive changes in beliefs about nature' which transform both the theory and practice of a science in a profound way. They are discontinuous transitions in which one mode of conceptualizing and interacting with the world is rejected and replaced by another incompatible rival. As such, their effects are both far-reaching and deep-rooted: an entire field is reconstructed from new fundamentals. That granted, we need to ask how rapidly these transitions take place, and whether or not Kuhn thinks that they are rationally intelligible.

The text that follows immediately on the quotation just given suggests that Kuhn believes that discontinuous transitions take time to occur. He writes that 'During the *transition period* there will be a large but never complete overlap between the problems that can be solved by the old and by the new paradigm. But there will also be a decisive difference in the modes of solution. *When the transition is complete*, the profession will have changed its view of the field, its methods, and its goals'.[25] However, Kuhn then alludes to Butterfield's description of the transition from Aristotelian physics to classical mechanics, and to Hanson's excursions into gestalt psychology, both of which suggest that revolutionary changes occur more or less instantaneously. As I pointed out above, Butterfield's argument can be reconciled with the view that it takes time to replace one system with another. Hanson's illustrative examples lead one to believe that replacement is immediate, however.

Hanson[26] suggested that after a revolutionary change people saw the world differently. He compared the transition across a revolutionary divide to the perceptual flips experienced when one sees a particular set of marks on paper now as an antelope, then as a bird. Other familiar instances of such gestalt switches are the Necker cube, which suddenly seems to change its orientation in space, and the so-called 'duck-rabbit'. In all such cases the *entire* image changes *instantaneously* — from a duck to a rabbit, for example.

Although he borrows the metaphor of a gestalt switch from Hanson, Kuhn introduces it with some hesitancy. It is, he says, 'a useful elementary prototype for what occurs in full-scale paradigm shift'.[27] He also remarks that when paradigm changes occur in science, they differ from gestalt switches in that they are 'usually more gradual and almost always irreversible'.[28] He is not entirely consistent, however. Crises in normal science, he notes, 'are terminated, not by deliberation and interpretation, but by a relatively sudden and unstructured event like the gestalt switch'.[29] Stunned as by a flash of lightening, scientists find that 'large portions' of their experience are transformed in one fell swoop. On this view, then, time almost shrinks to an instant, and in that instant a new intellectual framework, often incommensurable with its predecessor, is 'laid down . . . . on nature whole'[30] as Kuhn puts it.

Kuhn's uncertainty over the rate at which a discontinuous transition takes place reaches acute proportions in his seemingly incoherent assertion that 'like the gestalt switch, it must occur all at once (though not necessarily in an instant) or not at all'.[31] Plainly Kuhn feels that the gestalt metaphor has its limitations. Yet he is loath to abandon it, particularly because it performs another very important function in his theory of revolutionary change. It does not simply suggest that such changes are relatively sudden and comprehensive; it also indicates that ultimately they are not rationally intelligible.

As we have just seen, Kuhn insists that crises in normal science are not resolved by 'deliberation and interpretation', but by an event akin to a gestalt switch. His invocation of the metaphor, here and elsewhere in his work, reflects his conviction that there are limits to the value of argument or reasoning as motors of historical development. Resistance to change is inevitable and legitimate, he says, and is overcome — in the last instance — by what he calls a conversion experience. Kuhn does not deny that good reasons can be advanced by the proponents of a new theory in defence of their position. These provide motives for conversion and a climate in which it is more likely to occur. On their own, however, they are not sufficient to effect a change of allegiance. The discontinuous transition between competing paradigms, Kuhn writes, 'cannot be made a step at a time, forced by logic and neutral experience'.[32] It involves a relatively sudden and explosive revelation, a new way of looking at things, and the mechanism by which it occurs ultimately defies analysis. This means that 'communication across the revolutionary divide is inevitably partial . . . . the proponents of competing paradigms practice their trades in different worlds . . . . before they can hope to communicate fully, one group or another must experience the conversion that we have been calling a paradigm shift'.[33]

To summarize: on the basis of his historical analyses, Kuhn came to the conclusion that the advance of science has sometimes involved the destruction of one set of beliefs about the world and its replacement with another, and that such discontinuous transitions cannot be achieved by reasoning or argument alone. Effectively identifying the limits of reasoning with the limits of reason itself, Kuhn could not but conclude that the non-rational plays an important, even crucial, role in precipitating fundamental changes in science. In the notion of a gestalt switch he discovered the metaphor, which — despite its limitations — illustrated what he took to be the central features of revolutionary change. For Kuhn such changes are more or less sudden, more or less comprehensive and, in the last resort, non-rational transferrals of allegiance from one conceptual framework to another. All of these ideas are fused in the notion of the gestalt switch. As such it encapsulates the central features of what I am calling the liberal conception of revolution. What is more, its use reflects the espousal of that liberal conception of rationality which identifies reason with reasoning.

Kuhn's insistence that 'non-rational' considerations (i.e., for him, factors other than logic and experience) have played a role in shaping the growth of scientific knowledge, has disturbed many philosophers and historians. To argue that major scientific developments ultimately lie beyond the bounds of rational intelligibility is to question the rationality of science itself, and to undermine one of the central pillars on which its epistemological and social authority rests. It is also to imply that, in the last analysis, the nature and direction of scientific change cannot be controlled. By suggesting that the post-revolutionary order suddenly emerges almost full-blown in a somewhat arbitrary and random manner, 'gestaltism' undermines the efforts of those who would consciously intervene in the historical process. In the last resort, historical 'agents' simply become the victims of inscrutable forces and pressures that lie beyond their rational control.

Faced with these rather unpalatable conclusions, those who share Kuhn's liberal conception of revolution and of rationality, tend simply to reaffirm the view which he is challenging. Thus we find an eminent historian of the seventeenth century, Rupert Hall, adopting more or less as a working hypothesis the view that

> We may properly maintain in relation to the past that the process of the succession of theories in science was neither irrational nor random: that it was not largely directed by the workings of the unconscious, but by the conscious, reasoning mind; and that it occurred as the consequence of rational discourse between men, founded upon the knowledge of natural phenomena accessible to them.[34]

In similar vein we find Toulmin attempting to turn Kuhn, the historian of the Copernican Revolution, against Kuhn, the social psychologist. Kuhn's own work, says Toulmin, shows that 'If the men of the sixteenth and seventeenth centuries changed their minds about the structure of the planetary system, they were not forced, motivated, or cajoled into doing so; they were given reasons to do so. In a word, they did not have to be converted to Copernican astronomy; the arguments were there to convince them'.[35]

Now, as I pointed out earlier, Kuhn does not deny that there were reasons for espousing Copernicanism in the sixteenth and seventeenth centuries. What he does deny is that 'rational discourse between men', as Hall puts it, was sufficient to get its opponents to espouse the new world system. It is one thing to say that people were given reasons for becoming Copernicans, and that arguments were put forward to convince them. But whether or not those reasons or arguments were *good* ones, and even if they were, whether or not people were *in fact* persuaded by them: these are quite different matters. Toulmin's failure to perceive these distinctions amounts to an implicit assumption on his part that the arguments put forward by the Copernicans were so compelling that no sane person could possibly dissent from them. Effectively, then, his position is precisely that which Hall adopts as a working hypothesis, in other words that progress in science is directed by the 'conscious, reasoning mind'.

It is to Kuhn's credit that he refuses to make this unfounded assumption. Insisting that there have been discontinuous transitions in science involving the destruction of an established conceptual framework, he regards resistance to the new as inevitable and legitimate. On the other hand, while avoiding his critics' mistake, he promptly falls into the opposite trap. He shares his opponents' unduly restrictive concept of rationality and a liberal conception of revolution, that sees the new order as quickly replacing the old. Thus, as an alternative or complement to argument, he stresses the importance for historical development of relatively sudden, comprehensive and non-rational processes like a conversion experience or gestalt switch. It is because Hall and Toulmin, justifiably, retreat from this abyss that they, unjustifiably, place argument and discussion centre-stage as motors of historical development. Correlatively, Toulmin argues that although the development of science reveals conceptual discontinuities, they are neither as drastic nor as infrequent as Kuhn suggests, and that the progress of science is evolutionary rather than revolutionary.[36]

I remarked at the start of this section that the tendency to think of scientific and political changes in the same terms has been a feature of western thought since about the middle of the eighteenth century — and indeed we find connections of this kind being explicitly made by Toulmin. Having argued that Kuhnian-type revolutions have not occurred in science, he goes on to weave Kuhnian categories into the context of political theory with a view to showing that they are inappropriate there too. Liberal democratic thinkers, he says, used to believe that 'steady constitutional change represented a "rationally intelligible" political continuity; by contrast, political revolutions were disruptions of "normality" which introduced historical discontinuities unanalysable in normal, rational terms'. Now they know better; experience has shown that 'even the most unconstitutional change does not involve absolute and comprehensive breaches of political continuity'.[37] Any realistic theory of scientific and of political change must then be evolutionary. It will begin from the premise that revolutions are illusory and will recognize that science (society) comprizes an ' "historical population" of logically independent concepts and theories' (institutions) which are necessarily modified 'piecemeal or one at a time'.[38]

A theory of history which stresses continuity and piecemeal reform is also tied up with a liberal political stance through the conception of rationality which Kuhn, Hall and Toulmin all share. It is that conception which identifies rational change with reasoning and argument, and which takes violence (against, but not by, those who are willing to go on talking) as its direct antithesis. 'In the great tradition of Western rationalism', writes Popper, we 'fight our battles with words rather than with swords'.[39] This is unduly restrictive, of course. The law attempts to change people's behaviour, not by arguing with them, but by imposing sanctions for violating its prescriptions. Only in extreme and regrettable circumstances do these sanctions actually involve the death of the offender. Be that as it may, having espoused this dichotomy, it is but a short step to aligning discontinuity and revolution with violence and irrationality, opposing to it continuity, piecemeal reform and what Hall calls the rational discourse between men.

From the above it emerges that, once one is committed to a

liberal theory of revolution and of reason, there are strong pressures on one to adopt continuity as a methodological postulate in historical analyses. Within this framework to opt for discontinuity is to opt for sudden, possibly violent and cataclysmic destruction and replacement: it is to side with unreason. Fundamental, wide-ranging transformations are taken to be both earth-shattering and, ultimately, arbitrary, and the possibility of rationally controlling history is jeopardized. Faced with this alternative the advantages of a continuist position are many. It rescues historical change from the sway of blind forces, it locates the motor of progress in the conscious, reasoning mind, and it buttresses a political commitment to piecemeal reform.

Yet for all their apparent advantages the liberal conceptions have serious drawbacks, not least of all as far as an event like the Scientific Revolution is concerned. For, as we have seen, even some of those who adopt continuity as a methodological postulate, admit that this event involved a radical, dramatic and far-reaching transformation of the human understanding. However, within the liberal framework, to accept that the change was discontinuous is to claim that it was relatively sudden and, in the last resort, rationally unintelligible. On the other hand, to preserve its rationality is to deny its discontinuous and revolutionary character. Clearly, an alternative approach, free from the liberal assumptions that shape methodological continuity, is called for. It is my contention that the seeds of such an approach lie in the work of Koyré and, more recently, Clavelin. It is to them that I now turn.

1.2 The 'Marxist' Conception of Revolution

Koyré is one historian of science who has uncompromizingly faced up to the fact that Aristotelian physics was rejected *in toto* and was replaced by another system of thought. As I pointed out earlier, he believes that Galileo and his heirs 'had to destroy one world and replace it by another'. This meant that they had 'to reshape the framework of our intellect itself, to restate and to reform its concepts'. As far as Koyré is concerned, then, a discontinuous transition occurred in science in the seventeenth century involving a transformation of the human understanding and the production of a new conceptual framework incompatible with the orthodoxy of the day.

To appreciate the extent of Koyré's achievement it is imperative that one recognize that one of his central aims is to explore the processes whereby this new conceptual apparatus was produced. Thus, isolating the principle of inertia as one of the identifying characteristics of classical mechanics, he writes

> The principle of inertia did not emerge straight away in final form, like Athena from the head of Zeus, from the thought of Descartes or Galileo. The new concept of motion, with its implication of a new conception of physical reality, of which the princple of inertia is both the basis and the expression, was elaborated increasingly accurately as the result of a long and difficult mental labour. The Galilean and Cartesian revolution — for it was nonetheless a revolution — had a prolonged preparation. It is the history of this preparation which we intend to investigate here, a history which constitutes a necessary prerequisite for any understanding of Galileo's achievements, a history in which the human mind can be observed obstinately grappling over and over again with the same problems, untiringly coming up against the same objections, the same difficulties, and slowly and laboriously forging the instrument which would enable it to overcome them.[40]

Koyré's view that the transition to classical mechanics occurred 'slowly and laboriously' contrasts starkly with Kuhn's retreat to gestalt switches and conversion experiences to account for discontinuities. This difference in conceptualization is not accidental. It reflects their quite different ways of thinking about revolutionary change. For Koyré, revolution is a protracted and painful *process* of rejection and replacement. For Kuhn it is, in the last analysis, a more or less explosive and unstructured *event*, which destroys the old and replaces it with the new. In other words, Koyré sets out to explain the *process of production of new knowledges*. Kuhn, in line with Popper, Lakatos and, to a certain extent, Feyerabend, tends to define his task as accounting for *changes of allegiance from one more or less ready made knowledge-product to another*.

There is another way of looking at this. As I stressed earlier, Kuhn formulated his 'theory' of discontinuous transitions in opposition to the view that historical change is a piecemeal process in which development occurs by the successive accumulation of small increments. Along with Butterfield, he opposed the so-called Whig interpretation of history, which subordinates the past to the present, emphasizing likenesses, anticipations and roots, and generating a spurious continuity. And, with Butterfield, he agreed that the proper task of the historian was to understand the past for its own sake. For Kuhn this means that the opinions of historical agents must be located in their proper historical context, and studied 'from the viewpoint — usually very different from that of modern science — that gives those opinions the maximum internal coherence and the closest possible fit to nature'.[41]

What does this project amount to in Kuhn's case ? Kuhn in effect rationally reconstructs different conceptual systems that have emerged in the past, showing how their several elements are related to one another. In doing so, he treats them as fully articulated and relatively coherent bodies of thought. What he fails to appreciate is that it takes time to build up structures of the kind that he reconstructs. Thus Kuhn can write that 'to make the transition to Einstein's universe, the whole conceptual web whose strands are space, time, matter, force, and so on, had to be shifted and laid down again on nature whole'.[42] In other words Kuhn looks back on history from the vantage point of what is in fact the *outcome* of a protracted historical process. His is essentially a spectator's, not a participant's, perspective, and as such it completely misses the point than an immense intellectual effort is required to produce and extend a new conceptual scheme. His theory of revolutionary change is accordingly flawed; it is, in the last analysis, an ahistorical theory of gestalt switches.

Koyré and, more recently, Clavelin also rationally reconstruct now defunct conceptual frameworks like Aristotelianism. However, their aim in doing so is not to study the past for its own sake. On the other hand, neither are they interested in resurrecting the Peripatetic system. Unlike Feyerabend, for example, they do not explore the past with a view to demolishing the arrogance of those who think that modern science is the only intelligible way of making sense of the world. For Koyré and for Clavelin the cognitive respectability of Aristotelian thought is not in doubt. They rationally reconstruct it because it was as a 2000-year old richly developed system of thought that Galileo encountered it, and against which he rebelled. It is by doing so, by adopting an actor's perspective on the historical process, that they have been led to conclude that the development-by-accumulation thesis of the continuist historian is seriously defective.

Koyré's and Clavelin's approach to the historical record exemplifies some of the features of what Foucault calls an 'archaeological description' of what people have said or, more technically, of discursive practices. Such analyses also reject that tradition in the history of ideas that attempts to iron out differences by appeals to traditions, influences, anticipations and so on. Archaeology does not aim to dissolve differences, thereby retrieving the continuous. It

> seeks rather to untie all those knots that historians have patiently tied; it increases differences, blurs the lines of communication, and tries to make it more difficult to pass from one thing to another.... For the history of ideas, the appearance of difference indicates an error or a trap; instead of examining it, the clever historian must try to reduce it: to find beneath it a smaller difference, and beneath that an even smaller one, and so on until he reaches the ideal limit, the non-difference of perfect continuity. Archaeology, on the other hand, takes as the object of its description what is usually regarded as an obstacle: its aim is not to overcome differences, but to analyse them, to say what exactly they consist of, to *differentiate* them.[43]

This is indeed Koyré's starting point. It is the differences between classical physics and Aristotelian dynamics that interest him. In this spirit he attaches particular importance to the formulation of the principle of inertia. However, Koyré is not satisfied merely to record this difference. He wants 'to analyse it, to say exactly what it consists of'. Thus, while agreeing with those who characterize classical physics in this way, he goes on to say that nevertheless

> this characterisation seems somewhat superficial. It is not enough simply to point to these facts. It must be explained why modern physics has been able to adopt the principle of inertia i.e., it must be explained why and how this idea, which seems to us so very obvious, has been able to acquire this *status* as an *a priori*, self-evident truth, whereas for the Greeks and for medieval thought it seemed, on the contrary, to suffer from a self-evident and irremediable absurdity.[44]

A crucial aspect of Koyré's approach to revolution is beginning to emerge here. As far as he is concerned, the absence of the principle of inertia from Greek and medieval thought is not merely accidental. Modern physics, he writes, 'has been able' to formulate this principle; Aristotelian thought not only *did not*, it actually *could not* do so. Koyré goes on to explain why it was *impossible* for the Peripatetics to take this step, when he says that

> I believe that the intellectual attitude of classical science can be characterised by the following two changes, which are moreover intimately related: geometrisation of space and dissolution of the Cosmos, that is to say the disappearance from within scientific reasoning of the Cosmos as a presupposition and the substitution for the concrete space of pre-Galilean physics of the abstract space of Euclidean geometry. It was this substitution that made the invention of the law of inertia possible.[45]

For Koyré, then, it was essential that some of the fundamental assumptions of Aristotelianism be eliminated if the principle of inertia was to be formulated. A conceptual revolution was necessary in which these assumptions were jettisoned and replaced by others, which provided the material conditions under which it was possible to produce inertial physics. The reasoning lying behind this position is complex, and is explored at some length in Chapters 5 and 6 of this book. Nevertheless, it is instructive to illustrate the general point that he is making by briefly touching on one or two aspects of his argument.

As we have seen, the mathematization or geometrization of nature, and of science, is regarded by Koyré to be one of the two *specific* characteristics of the Scientific Revolution. The other, complementary, feature was the destruction of the Cosmos and the associated hierarchical conception of the universe. The *absence* of mathematization in Aristotelian thought is thus of particular significance for him. It is to be treated as a symptom of certain constraints built into the framework of Aristotelianism itself, which excluded it. It could only rise to prominence, therefore, along with an entirely different framework; piecemeal reform of the prevailing system of thought would never have ensured its survival.

Cassirer and Randall effectively deny this. Schmitt, in turn, has persuasively defended Koyré's position. By isolating the main threads of this debate, we can gain some insight into the conception of revolution which is implicit in Koyré's work.

The argument focusses on the work of Iacopo Zabarella (1553–1589) of the School of Padua. Both Cassirer and Randall see in it the culmination of a long, incremental development of

Aristotelian methods, and an anticipation of the Galilean approach. Randall argues that Zabarella's originality, and that of the tradition which he summed up, lay in the distinction that he drew between 'scientific experience' on the one hand, and the random or planless collection of observations on the other. Zabarella did not use mathematics, however; according to Randall, that 'gradually emerged' into its dominant position as a result of the activities of a small group of initially marginal scholars. Randall continues that 'With this mathematical emphasis added to the logical methodology of Zabarella, there stands completed the 'new method' for which men had been so eagerly seeking . . . . it turned out to be the least novel of all the elements that went into the formation of the new science'.[46]

Randall's claim is instructive in that it reveals two of the characteristic features of a continuist historiography, viz. the piecemeal, cumulative conception of development and the denial of originality. The so-called new method of Galilean science, he claims, can be obtained simply by adding a mathematical emphasis to Zabarella's logical methodology. Koyré, while describing Randall's article as brilliant, also claims that Randall's defence of continuity is incompatible with his admission that Zabarella effectively ignored the importance of mathematics in natural science. Schmitt explains that absence in the following terms

> The major difference which we find between the two thinkers seems to stem from the entirely different ways in which they viewed the world. Zabarella, following Aristotle and the long Aristotelian tradition, saw it as a living, biological entity, teleologically oriented and best understood through experience and syllogistic reasoning. Galileo, on the other hand, saw the world in mathematical terms, composed of geometrical figures which move in conformity with mathematically expressible laws and best understood through the modes of mathematical analysis. It would be difficult to overemphasize the significance of this distinction when one is faced with the evolution of seventeenth-century science.[47]

In short, against Randall, Schmitt argues that one cannot consistently hold that Zabarella failed to appreciate the importance of mathematics *and* that all that Galileo had to do was to 'add' this missing component to the Aristotelian tradition. The new scientific method developed in the seventeenth century was genuinely novel, and its absence from Zabarella's work is not accidental.

Schmitt's point then, somewhat reformulated, is that, in order to make sense of certain natural phenomena (e.g., freely falling bodies), the Aristotelians developed a framework which pivoted on a number of metaphysical assumptions about the nature of reality. On that basis they generated a set of concepts for explaining change, or motion in general, and for explaining the motion of heavy bodies on earth, in particular. Because they conceived of any and every change, including locomotion, in qualitative, teleological terms, mathematics was only of secondary importance to them for developing a physics of motion. Detailed attention to it, and the emergence of experimental methods that relied on it, were excluded by the presuppositions of the Aristotelian system of thought. Those presuppositions *imposed limits* on the kinds of questions one posed as a committed Aristotelian, the kinds of issues one thought were worth pursuing, the directions in which one believed progress lay. Galileo could not therefore simply 'add' a mathematical approach to the prevailing Aristotelian methodology when developing his new physics. He had to reject Aristotelianism along with its fundamental metaphysical assumptions. To repeat Koyré's now familiar claim, Galileo had to reshape our intellect itself; 'to give to it a series of new concepts, to evolve a new approach to being, a new concept of nature, a new concept of science, in other words, a new philosophy'.[48]

Yet Galileo was not a Newton. Although he rebelled openly against Peripatetic thought and the institutions that embodied it, he was not able completely to sever his links with it. As far as Koyré and Clavelin are concerned, Galileo never managed to formulate the 'classical' principle of inertia. For Galileo it was circular, not rectilinear, motion that was conserved. Constrained by his Greek heritage, he could not fully transcend the limits that it imposed on the process of conceptual innovation that he undertook. 'The impossibility of Galileo's arriving at the principle of inertia', writes Koyré

is to be explained on the one hand by his refusal to completely abandon the idea of the Cosmos, i.e., the idea of a well-ordered world, and to unambiguously accept the infinity of space; and, on the other hand, by his inability to conceive of physical bodies (or the bodies of physics) as being without the constitutive property of gravity.[49]

As I noted above, the former idea was a fundamental tenet of Aristotelianism. The view that gravity was an inherent property of bodies is a spontaneous, commonsensical way of conceiving of it, which Koyré traces to Aristotelian and Archimedean thought.

Galileo, then, is a transitional figure. His break with his past was neither clean nor complete. As Clavelin puts it,

> though Galileo succeeded in drawing conclusions that classical dynamics would fully endorse, he did so by means of a conceptual system that in no way foreshadowed that of classical dynamics . . . . Galileo could do no better than to erect a scaffolding, one that was bound to be removed just as soon as the new edifice had been completed.[50]

This aspect of revolutionary change has been ably stressed by Marx, who writes

> Men make their own history, but they do not make it just as they please; they do not make it under circumstances chosen by themselves, but under circumstances directly encountered, given and transmitted from the past. The tradition of all the dead generations weighs like a nightmare on the brain of the living. And just when they seem engaged in revolutionising themselves and things, in creating something that has never yet existed, precisely in such periods of revolutionary crisis they anxiously conjure up the spirits of the past to their service and borrow from them names, battle cries and costumes in order to present the new scene of world history in this time-honoured disguise and this borrowed language.[51]

In similar vein Ibsen has one of his heroines, Mrs Alving, say

> It isn't just what we have inherited from our father and mother that walks in us. It is all kinds of dead ideas and all sorts of old and obsolete beliefs. They are not alive in us; but they remain in us none the less and we can never rid ourselves of them. I only have to take a newspaper and read it, and I see ghosts between the lines. There must be ghosts all over the country. They lie as thick as grains of sand. And we're all so horribly afraid of the light.[52]

These 'ghosts' surreptitiously undermine the attempts made by a new system to forge its own identity, and subvert its struggles to distance itself from the structure to which it is opposed. Yet even while acting as obstacles to its development, they play an essential role in the construction of a different pattern of relationships. As Clavelin notes, Aristotelianism was 'the conceptual universe from which (and also in opposition to which) classical mechanics was founded'.[53] It is not surprising that it should have lived on fragmentarily in Galileo's thought, circumscribing the scope of his achievement. Several of his principles and concepts may have been somewhat rough and ready. At times his arguments may have been contradictory and even regressive. This is only to be expected. For, writes Clavelin

> The uncertainties and limitations in his path were the typical obstacles encountered by every creative thinker who finds that each new principle and each new concept must be wrested from diametrically opposite principles and concepts and that each new step not only introduces a new content but also marks an advance from one intellectual universe to another.[54]

The broad outline of the conception of revolution, which is implicit in these analyses, has now emerged. Both Koyré and Clavelin begin by accepting the fact that Aristotelian cosmology and mechanics were rejected by the natural philosophers of the seventeenth century, and replaced with a new system of thought. However, in contrast to Kuhn's theory, they emphasize that that transition took time — it was a process not an event — and that it involved an almost superhuman intellectual effort. What's more, they recognize that there is a double aspect to revolutionary transformations. On the one hand, a new pattern of relationships is gradually built up *in opposition to* the old order, which it ultimately replaces. When considered across a sufficiently wide timespan, such transitions are thus seen to be discontinuous. On the other hand, the new system also *emerges from* the old order and carries with it, particularly in its early stages, survivals and residues that betray its heritage. It is these that give a spurious appearance of continuity to what are in fact revolutionary changes.

It emerges from the analyses of these two authors that Galileo's rejection of Aristotelianism amounted to the recognition that the solution of certain problems which had become focal points of controversy since Copernicus' *De Revolutionibus* not only had not, but actually *could not* be solved within the Peripatetic framework. Having grasped this, Galileo set out to lay the foundations of a new physics within the womb of the old, struggling both intellectually and politically against the prevailing orthodoxy. These foundations were built on by his successors who gradually managed to eliminate the outworn components of the older order in the new, differentiating it ever more thoroughly from its crumbling rival, and transforming the human understanding in the process. The outcome of this achievement was that magnificent structure that we call classical mechanics — a system which was in turn to become an orthodoxy against which a rather different kind of rebellion recently took place.

Methodologically speaking, this theory of history aims to isolate the specificity of an event like the Scientific Revolution and is ever alert for those of its aspects which differentiate it from that which came before. Looking back from the vantage point of the achievements of a Newton, it postulates that their previous absences were not accidental, but that they were structurally excluded by the system of thought that was ultimately eliminated. It thus shows that the progress that they represent was only possible by virtue of a rupture with the past, which ultimately involved a transformation of the understanding.

Yet even while insisting on discontinuity, this theory recognizes the importance of continuity; its site is those residues and survivals of the old order that live on in the new. The 'Marxist' theory of revolution thus locates Galileo in the historical stream of his day; he is not seen as being above history, leaping arbitrarily from one world view to another. Rather, he is always constrained by his present, drawing on it even as he struggles against its stifling grip. Discontinuity of this kind does not deny continuity; it accepts it. Roughly speaking, it shares with Braudel the view that

> mankind is more than waist deep in daily routine, countless inherited acts accumulated pell-mell and repeated time after time to this very day, that become habits that help us to live, imprison us and make decisions for us throughout our lives. Ancient, yet still alive, this multi-centuried past flows into the present like the Amazon pouring into the Atlantic ocean the vast flood of its cloudy waters.[55]

What it emphasizes is that through transformations of the understanding achieved by determined opposition and criticism of the common-sense of the day — the habits that imprison us and make decisions for us — those waters are slowly becoming clearer, and 'mankind' is gradually becoming more able to make its history rather than being made by it.

It must be emphasized that the 'Marxist' conception of discontinuous change outlined above is compatible with there being a 'moment' of revolutionary rupture. However, this moment is one in which the seeds of a new and ultimately victorious conceptual framework are sown. It is not the moment of a gestalt switch for, as I have stressed, the replacement of one system by another is a protracted and laborious process. The 'Marxist' theory of revolutionary change is thus a theory of levels moving at different speeds, a theory which sees the new as struggling to differentiate itself from the old, from which it breaks and in which it is initially embedded.

At the beginning of this chapter I remarked that a concept of revolution inspired by the French and American revolutions, had been frequently used since the late eighteenth century to describe major innovations in science. A notable feature of it was that it took revolutions to be both sudden and far-reaching; I quoted Berthelot as saying that, in the French Revolution, society was reconstituted on new foundations and a basically new era established in just fifteen years. This is the 'liberal' conception of revolution which 'gestaltism' pushes to an extreme limit, shrinking time to an instant and the transformation achieved in that instant to a universe.

However, since the Russian Revolution, a new concept of revolution has risen to prominence. Cohen, in fact, points out that scholars today have adopted the notion of a continuing revolution, which he takes to be one of the innovative features of the Russian experience.[56] To indicate that the Revolution needed to be *consolidated* and that this would take time, the Russian revolutionaries reckoned the calendar in years of, not since, the Revolution. The idea that revolutions do not occur within a short space of time, says Cohen, is now generally accepted among historians of science, and is reflected in one of Hall's books, *The Scientific Revolution, 1500–1800*. This immediately suggests the possibility of there being a fruitful exchange between Marxist political theory, on the one hand, and the history of science, on the other, both being concerned to develop an adequate theory of discontinuous transition.

I shall give just two quotations to illustrate the resonances. The first stresses the discontinuous component in a revolutionary transformation. Thus Colletti points out that, for Lenin

> the destruction of the old [state] machine is the destruction of the *limits* imposed on democracy by the bourgeois state. It is the passage from a 'narrow restricted' democracy to a full democracy. And, adds Lenin, 'full democracy is *not*, qualitatively, the same thing as incomplete democracy'. Behind what might seem formally a difference in quantity what is actually at stake is 'a gigantic replacement of certain institutions by other institutions of a fundamentally different type'.[57]

For Lenin, then, a revolutionary change involves the replacement of one set of institutions which impose limits on democracy by a qualitatively different, more adequately democratic network of organizations.

The second quotation alludes to the continuity that is also a characteristic of the 'Marxist' theory of revolutionary change. It derives from a comment by Fagen on the claim made in a Cuban document to the effect that 'the past has its claws into the present'. Fagen writes that the 'inertia of a cultural system is very great indeed'. The Cuban revolutionary leaders, he says, have had to learn to live with 'some of the ideas, habits and concepts that were well integrated into the pre-revolutionary way of life . . . . The developmental effort is constantly and inevitably slowed down by the persistence of old values and patterns of behaviour, even in the bosom of the newest and most disciplined of revolutionary institutions'.[58]

Once we reject the idea that revolutions are necessarily sudden, cataclysmic, ultimately non-rational events, we need also to revise our conception of what constitutes rational progress in science and in society. The liberal conception of reason, as I pointed out above, identifies reason with reasoning, and takes argument to be the rational motor of historical development. Criticism (and thus growth) is effected through the notions of contradiction, consistency, implication and so forth. These relationships are thought to hold essentially between propositions, items that can be true or false and are the content of our statements and arguments. Mediated by these concepts reason exerts its critical, progressive force only through reasoning and discussion. Hall, it may be remembered, eloquently expressed the associated theory of history when he asserted that scientific progress is largely directed 'by the conscious reasoning mind', and that it occurs 'as the consequence of rational discourse between men'.

The attempts that have been made recently, particularly by Edgley,[59] to develop a dialectical conception of reason suggest that it is in this direction that the articulation of an adequate theory of rational progress lies. Very briefly it emphasizes the importance of criticism and opposition, and particularly of contradiction, as a motor of progress. However, it does not see reason as being critical simply of arguments, but of beliefs and actions. It thus transcends that arid conception of history, which as Dunn puts it, simply

> maps the logic of arguments and sets these out against its own prescriptive logic, so that their structure can be grasped clearly. All the statements contained in it are statements about the relationships of propositions to propositions. Men, breathing, excreting, hating, mocking, never step inside it. Their role is merely to label a particular set of propositions with the name which they bear themselves. Their names appear in this story but never their selves. It is a tale to be told by clever and subtle men, and it signifies much but in it there is neither sound nor fury.[60]

By locating the beliefs and actions of historical agents as targets for criticism and opposition, dialectical reason effectively takes a participant's perspective on the process of change. And it recognizes that behind the 'conscious reasoning mind', mocking its arrogance and pretentiousness, lie the habits of thought and action which imprison us and make decisions for us. It is thus, potentially, a far richer conception of rationality than one that restricts reasonable means of achieving progress to critical discussion. Discontinuous transitions are as wide as they are deep. To overcome resistance to them procedures other than argument or reasoning may be required, and can be defended as rational.

1.3 Conclusion

My aim in this chapter has been to demarcate a terrain, to establish landmarks and points of reference. For reasons that will by now have become clear, the related concepts of discontinuity and incommensurability are as important as they are controversial. At present in most, if not all, debates in the philosophy of science the Kuhnian version of the thesis is dominant. However, given the serious flaws that it has, Kuhn's 'theory' of discontinuous transition is more or less readily dismissed. Sweet reason calmly and triumphantly disposes of it. This chapter will have achieved one of its major objectives if it succeeds in dispelling some of the more or less standard objections to discontinuity, and if it succeeds in re-opening the problem of ruptures and breaks in history on ground different to that occupied by Kuhn.

As I pointed out in the Introduction, Kuhn has said that both he and his Popperian opponents share an interest in revolutionary change. Given the limitations of Kuhn's approach, it is instructive to explore the views of some of his well known colleagues and critics, with particular reference to their conception of revolutionary change. It is in this spirit that I now turn to a consideration of Popper's theory of rational scientific growth.

Notes

1. A.C. Crombie, *Augustine to Galileo* (Heinemann, London, 1952), p. xii.
2. *Ibid*. p.278.
3. A.R. Hall, *The Scientific Revolution 1500–1800* (Longmans, London, 1954), pp. xiii–xiv.
4. H. Butterfield, *The Origins of Modern Science 1300–1800* (Bell, London, 1968), pp. vii–viii.
5. S. Drake, *Galileo Studies* (University of Michigan Press, Ann Arbor, 1970), p.19.
6. A. Koyré, *Metaphysics and Measurement* (Chapman & Hall, London, 1968), p.21; hereafter cited as *Metaphysics*.
7. *Ibid*. pp.19-20.
8. S. Drake, *op. cit.*
9. P.P. Wiener and A. Noland (eds.), *Roots of Scientific Thought: A Critical Perspective* (Basic Books, New York, 1957), p.7.
10. E.J. Dijksterhuis, 'The origins of classical mechanics from Aristotle to Newton', in M. Clagett (ed), *Critical Problems in the History of Science* (University of Wisconsin Press, Madison, 1959), p.178.
11. A.F. Chalmers, *What is This Thing Called Science?* (University of Queensland Press, St. Lucia, 1976), p.xii.
12. I.B. Cohen, 'The eighteenth-century origins of the concept of scientific revolution', *Journal of the History of Ideas*, 37 (1976), 257–8.

13 *Ibid.* p.259.
14 M. Berthelot, quoted in L.S. Feuer, *Einstein and the Generations of Science* (Basic Books, New York, 1974), p.244.
15 M. Berthelot, quoted in L.S. Feuer, *Ibid.* p.245.
16 H. Butterfield, *op.cit.* p.7.
17 *Ibid.* p.5.
18 T.S. Kuhn, *The Structure of Scientific Revolutions* (University of Chicago Press, Chicago, second edition, 1970), pp.1–2.
19 H. Butterfield, *The Whig Interpretation of History* (Bell, London, 1931), pp. 11–12.
20 T.S. Kuhn, *op.cit.* pp.97–8.
21 *Ibid.* p.42.
22 T.S. Kuhn, 'Reflections on my critics', in I. Lakatos and A. Musgrave (eds.), *Criticism and the Growth of Knowledge* (Cambridge University Press, Cambridge, 1970), p.250.
23 T.S. Kuhn, *op.cit.* p.6.
24 *Ibid.* pp.84–5.
25 *Ibid.* p.85 (my italics).
26 N.R. Hanson, *Patterns of Discovery* (Cambridge University Press, Cambridge, 1958), Chapter 1.
27 T.S. Kuhn, *op.cit.* p.85.
28 *Ibid.* p.111.
29 *Ibid.* p.122.
30 *Ibid.* p.149.
31 *Ibid.* p.150.
32 *Ibid.* p.150.
33 *Ibid.* pp.149–150.
34 A.R. Hall, 'Magic, metaphysics and mysticism in the Scientific Revolution', in M. Righini Bonelli and W.R. Shea, (eds), *Reason Experiment and Mysticism in the Scientific Revolution* (Science History Publications, New York, 1975), pp.281–2.
35 S. Toulmin, *Human Understanding*, Vol. I (Clarendon Press, Oxford, 1972), p.105.
36 S. Toulmin, 'Does the distinction between normal and revolutionary science hold water ?', in I. Lakatos and A. Musgrave (eds), *op.cit.* pp.39–47.
37 S. Toulmin, *op.cit.* p.117.
38 *Ibid.* pp.129, 130.
39 K.R. Popper, *The Open Society and its Enemies*, Vol. II (Routledge & Kegan Paul, London, fifth edition, 1966), p.396.
40 A. Koyré, (trans. J. Mepham), *Galileo Studies* (Harvester Press, Hassocks, 1978), p.131.
41 T.S. Kuhn, *op.cit.* p.3.
42 *Ibid.* p.149.
43 M. Foucault, *The Archaeology of Knowledge* (Tavistock Publications, London, 1974), pp.170–71.
44 A. Koyré, *Galileo Studies, op. cit.* p.2.
45 *Ibid.* pp.2–3.
46 J.H. Randall, Jr, 'The development of scientific method in the school of

Padua', *Journal of the History of Ideas*, 1 (1940), 177, 205.
47 C.B. Schmitt, 'Experience and experiment: A comparison of Zabarella's view with Galileo's in *De Motu*', *Studies in the Renaissance*, XVI (1969), 124.
48 A. Koyré, *Metaphysics*, *op.cit.* p.3.
49 A. Koyré, *Galileo Studies*, *op.cit.* p.187.
50 M. Clavelin, (trans A.J. Pomerans), *The Natural Philosophy of Galileo* (MIT Press, Cambridge Mass., and London, 1974), p.357.
51 K. Marx, 'The eighteenth brumaire of Louis Bonaparte', in K. Marx and F. Engels, *Selected Works* (Lawrence & Wishart, London, 1970), p.96.
52 H. Ibsen, *Ghosts*, Act II.
53 M. Clavelin, *op.cit.* p.11.
54 *Ibid.* pp.264–5.
55 F. Braudel, quoted by P. Scott, *Times Higher Education Supplement*, 9 December 1977, p.9.
56 I.B. Cohen, *op.cit.* p.259.
57 L. Colletti, 'Lenin's *State and Revolution*', in R. Blackburn (ed), *Revolution and Class Struggle* (Fontana, London, 1977), p.71.
58 R.R. Fagen, 'A perspective on the transformation of political culture in Cuba', in S.A. Halper and J.R. Sterling, (eds), *Latin America: The Dynamics of Social Change* (Allison & Busby, London, 1972), p.195.
59 R. Edgley, 'Science, social science and socialist science: Reason as dialectic', *Radical Philosophy*, No. 15 (Autumn 1976), pp.2–7; also R. Edgley, 'The contradiction of Colletti', *Critique*, No.7 (Winter 1976/7), pp.47–52.
60 J. Dunn, 'The identity of the history of ideas', *Philosophy*, 43 (1968), 92.

2 THE POVERTY OF POPPERISM

THERE are a good many reasons for thinking that in Popper's work we have a potentially rich and fertile source from which to extract at least some of the basic components needed for a theory of discontinuous transition. An ardent champion of radical change in science, Popper has suggested that 'revolution in permanence' could just as well be the motto of scientific progress. According to him, Einstein's achievement should be a warning to us all: a warning that no scientific theory, no matter how well corroborated nor how deeply entrenched is beyond criticism, and that it can be overthrown by a superior rival. For Popper, the growth of scientific knowledge is not achieved by steadily accumulating observations. Rather, it consists in the 'repeated overthrow of scientific theories and their replacement by better or more satisfactory ones'.[1]

Here at the heart of his epistemology we appear to have an unambiguous commitment by Popper to discontinuity. Scientific advance, he says, involves the rejection and elimination of inadequate theories, and their replacement by more adequate alternatives. One of the major tasks which he has set himself is to develop a theory of rational scientific growth, which is compatible with this concern for revolutionary change. Regrettably, however, the different elements of that theory are scattered through Popper's work; to the best of my knowledge he has never linked them together into a coherent whole. This is doubly disadvantageous. On the one hand, it makes the task of interpretation both arduous and hazardous. Popper's theory of rational revolutionary progress has to be assembled from disparate sources, which were written with different purposes in mind. On the other hand, it means that Popper himself is often neither aware of the implications of ambiguities in his formulations, nor has he confronted and dealt with them.

To appreciate the point that I am making, consider the contradictory and confused stance that Popper adopts on the question of whether an entire system of thought, or only part of it, is wiped out by criticism. On the one hand, we find him writing

> It is necessary for us to see that of the two main ways in which we may explain the growth of science, one is rather unimportant and the other is important. The first [less important one — JK] explains science by the accumulation of knowledge: it is like a growing library (or a museum). As more and more books accumulate, so more and more knowledge accumulates. The other explains it by criticism: it grows by a method more revolutionary than accumulation — by a method which destroys, changes, and alters, the whole thing including its most important instrument, the language in which our myths and theories are formulated.[2]

Here Popper quite explicity states that in a revolutionary change the *whole* of our knowledge is destroyed, down to the very language itself in which our theories are formulated. This drastic claim is not easily reconciled with his insistence that, in discussing a scientific problem, the protagonists of different views have a common background knowledge the criticism of which 'must be *piecemeal*'. Popper explains why when he says that

> Almost all of the vast amount of background knowledge which we constantly use in any informal discussion will, for practical reasons, necessarily remain unquestioned; and the misguided attempt to question it all — that is to say, *to start from scratch* — can easily lead to the breakdown of a critical debate. (Were we to start the race where Adam started, I know of no reason why we should get any further than Adam did.)[3]

Sometimes Popper mixes these two views in an unholy alliance. In the following quotation, for example, he is discussing the rational response to a clash between observation and a theory or 'horizon of expectations' — a term which I shall elucidate later. As my emphases reveal, Popper begins by saying that an *entire* framework can be destroyed by recalcitrant evidence, but ends up by implying that knowledge grows by changing only *parts* of that horizon: he writes that observations

> can, under certain circumstances, *destroy even the frame itself*, if they clash with certain of the expectations. In such a case they can have an effect upon our horizon of expectations like a bombshell. This bombshell may force us to reconstruct, or rebuild, our *whole horizon* of expectations; that is to say, we may have to correct our expectations and fit them together again into

something like a consistent whole. We can say that in this way our horizon of expectations is raised to and reconstructed on a higher level, and that we reach in this way a new stage in the evolution of our experience; a stage in which those expectations *which have not been hit by the bomb* are somehow incorporated into the horizon, while those *parts of the horizon* which have suffered damage are repaired and rebuilt.[4]

The contradictions in Popper's position at this rather basic level of his work are also manifest in some of its less central aspects. For example, he has said that by and large the progress of science has not been 'slow, steady and continuous', involving 'successive degrees of improvement'.[5] On the other hand he has said of the Pre-socratics that 'there is the most perfect possible continuity of thought between their theories and the later developments in physics',[6] and he sees in Wegener's theory of continental drift a form of Thales' theory that the Earth floats on water !

Tensions of this kind have driven Lakatos, whose work I shall discuss at length in the next chapter, to despair. 'For Popper,' writes Lakatos, 'progress consists of an incessant, ruthless, revolutionary confrontation of bold, speculative theories and repeatable observations, and of the subsequent fast elimination of the defeated theories'.[7] Lakatos believes that this strand in Popper's thought destroys continuity in history by insisting on rapid theory *elimination*. In fact the alternative that he develops, his so-called methodology of scientific research programmes is specifically put forward by Lakatos to satisfy 'the requirement of continuous growth' in science.[8] As far as he is concerned, however, the seeds of this alternative also lie in Popper's work: he sees his own contribution as merely a 'small development' of one aspect of his mentor's project. Subsequently, however, he has admitted that he has had the greatest difficulty in unravelling the tangled web of Popper's views and that he was 'now at a loss as an exegetic'.[9]

In the light of the above, it would be misleading to attribute a particular theory of rational revolutionary progress to Popper without admitting that one has to stretch a point here, and to overlook an ambiguous or even contradictory comment there. Be that as it may, I think that a more or less coherent view can be built up which is compatible with the spirit if not always with the letter of Popper's position. In what follows I shall argue that the essence of his position is that spelt out in Popper's recent remarks that progress in science 'although revolutionary rather than merely cumulative, is in a certain sense always conservative: a new theory, however revolutionary, must always be able to explain fully the success of its predecessor',[10] and that, to be rational, 'a scientific revolution, however radical, cannot really break with tradition, since it must preserve the success of its predecessors. This is why scientific revolutions are rational'.[11] In other words it is my contention that one of the overriding aims of Popper's epistemology is to theorize Einstein's claim, which he displays so prominently, that 'there could be no fairer destiny for any . . . . theory than that it should point the way to a more comprehensive theory in which it lives on, as a limiting case'.[12] For Popper, 'the advance of knowledge consists, mainly, in the modification of earlier knowledge'.[13] My main aim in this chapter is to show how Popper's theory of revolutionary change is integrated into a framework of which this assertion is a central tenet.

2.1 Expectations and Problems, Conjectures and Refutations

We have seen that Popper makes apparently contradictory claims about how scientific progress is or ought (rationally) to be achieved. On the one hand, he says that knowledge advances mainly by *modifying* what is already known. On the other hand, he asserts that it grows by *replacing* defective theories with better or more satisfactory ones. To my mind, this contradiction is more apparent than real, however. To resolve it though one must distinguish carefully between the fortunes of well established theories and of novel conjectures in Popper's conception of rational change. My aim in this section is to sharpen this distinction and to begin to draw out some of its implications.

Popper's overall position can be conveniently presented using the tetradic schema $P_1 \rightarrow TT \rightarrow EE \rightarrow P_2$, which he elaborates in several ways.[14] Here P_1 is a problem from which we start. TT stands for a tentative theory — an imaginative and conjectural solution to P_1. Ideally several such solutions will be advanced. They are sifted at the stage EE, which refers to the process of error elimination. EE being successfully completed, a new situation arises which produces new problems of greater depth (P_2). Thereupon the cycle is repeated. Popper sometimes floats this scheme above what he calls 'background knowledge',[15] and his diagram suggests that this background knowledge is *not* affected by the transition from P_1 to P_2.

At the core of this schema lie those steps that are usually taken to be constitutive of Popper's epistemology. The essential features of what he rather blandly calls his method of trial and error are adequately spelt out in the following composite quotation

> The way in which our knowledge progresses, and especially our scientific knowledge, is by unjustified (and unjustifiable) anticipations, by guesses, by tentative solutions to our problems, by *conjectures*. These conjectures are controlled by criticism; that is by attempted *refutations*, which include severely critical tests . . . . If the outcome of a test shows that the theory is erroneous, then it is eliminated; the method of trial and error is essentially a method of elimination . . . . In this way we may, if we are lucky, secure the survival of the fittest theory by the elimination of those which are less fit.[16]

Now, as I have said, this procedure of conjecture, refutation (particularly in the light of recalcitrant evidence) and theory elimination seemingly leaves our background knowledge intact. And indeed Popper is emphatic that it would be fatal to destroy what we already know. As far as he is concerned

> In science it would be a tremendous loss if we were to say: 'We are not making very much progress. Let us sweep away all science and start afresh'. The rational procedure is to correct it and to revolutionize it, but not to sweep it away. You may create a new theory, but the new theory is created in order to solve those problems which the old theory did not solve.[17]

This remark has the merit of focussing attention on the role of problems in Popper's tetradic schema. It is now clear that, for him, problems are the points of intersection between new and existing theories. It is because problems arise with our existing theories, problems which trigger the creation of a hypothesis or a conjecture, that our knowledge progresses. Popper holds that these problems can arise in different ways and says that

> Science should be visualized as *progressing from problems to problems* — to

problems of ever increasing depth . . . . Problems crop up especially when we are disappointed in our expectations, or when our theories involve us in difficulties, in contradictions; and these may arise either within a theory, or between two different theories, or as the result of a clash between our theories and our observations . . . . Thus science starts from problems, and not from observations; though observations may give rise to a problem, especially if they are *unexpected*; that is to say, if they clash with our expectations or theories.[18]

Observations, then, can generate problems when they clash with theories or with our expectations. This is indeed a most unfortunate formulation. For sometimes when Popper speaks of theories he means conjectures; and for him, when a conjecture clashes with experience it is *refuted*. Does he want to suggest that this need not be so, that when conjectures or new, tentative hypotheses run into empirical difficulty a problem can be generated? It is my contention that he does not, and that to appreciate the precise role of problems in his epistemology is to gain an important insight into his theory of rational scientific change. To develop this line of argument it is necessary to look in a little more detail at what Popper means when he speaks of 'background knowledge' or of our 'expectations'.

Popper's use of a term like expectation does not signify some lapse of rigour on his part. It is an essential aspect of his solution to what he sees as the problem of induction, or 'Hume's problem' as he sometimes calls it. One of the basic facts noted by Hume was that reasonable people expect the future to resemble the past, assuming that events which they have not yet experienced will resemble those with which they are already familiar. These convictions are essential if purposive human action is to be possible, since such action, being future-oriented, presupposes that reality will behave in a more or less regular fashion. But according to Hume these convictions cannot be rationally justified. From this he drew the sceptical conclusion that in their everyday lives people were and needed to be illogical, and were guided by an irrational faith in the regularity of reality.

Popper insists that this scepticism is avoidable, and that our beliefs about the natural world can be rationally held, though not *inductively* justified. According to him, Kant was correct when he suggested that our intellect attempts to impose laws and regularities on the world. However, such expectations, while being psychologically *a priori*, in the sense that they are prior to experience, are not *a priori* valid, as Kant thought. In fact our expectations are frequently frustrated. But as long as we are prepared to adopt a critical attitude towards them, being willing to revise and to modify our pre-conceptions when they are disappointed, we are justified in (tentatively) holding them. What's more, they provide an indispensable framework for the growth of knowledge. For our minds are not 'buckets' into which are poured observational or perceptual raw material for scientific theorizing. Rather

> At every instant of our pre-scientific or scientific development we are living in the centre of what I usually call a *'horizon of expectations'* . . . . We may characterize an expectation as a *disposition to react, or as a preparation for a reaction*, which is adapted to or which anticipates a state of the environment yet to come about . . . . The horizon of expectations plays the part of a frame of reference: only their setting in this frame confers meaning or significance on our experiences, actions and observations.[19]

Earlier in this section I quoted Popper as saying that we should revolutionize our theories but not sweep them away. There is a conservative aspect to his conception of revolutionary change, and it now emerges that this is linked in Popper's mind with the idea that we live at the centre of a horizon of expectations. Thus

> Science never starts from scratch; it can never be described as free from assumptions; for at every instant it presupposes a horizon of expectations – yesterday's horizon of expectations, as it were. Today's science is built upon yesterday's science . . . . ; and yesterday's science, in turn, is based on the science of the day before. And the oldest scientific theories are built on pre-scientific myths, and these, in their turn, on still older expectations.[20]

A strong case can be made, then, for the view that when an existing, deeply entrenched system of thought clashes with experience it is not refuted and eliminated; rather, a *problem* is generated. Popper's epistemology *posits* a scientific community socialized into the traditions of its discipline, and it lays down procedures whereby that tradition may subsequently be *modified* in a reasonable way. These modifications are triggered when expectations are disappointed. This creates a problem for the scientific practitioner which is solved by the method of conjecture, refutation and theory elimination.

If my interpretation is correct, it follows that when discussing Popper's epistemology a careful distinction must be drawn between theories which are put forward as conjectures and theories which are constitutive of a horizon of expectations. The immediate consequences of these two types of theories clashing with experience are not the same. When an expectation clashes with experience a *problem* is produced. The appropriate response to it is a bold conjecture, springing from the imagination, from poetic intuition, from irresponsible dreams. But when a conjecture clashes with experience it is *refuted* and the appropriate response is to *eliminate* it and to put forward another tentative hypothesis in its stead. In other words, for Popper two quite different procedures are called for depending on whether an existing, well-entrenched system of thought clashes with experience or a novel, tentatively advanced theory does so. This is not to say that well-entrenched theories cannot be refuted by experience; in fact, as we shall see, Popper specifically allows for this possibility. What it does mean is that, in the absence of an alternative, the clash between such theories and observation creates a problem, not an immediate falsification. It also means that, though refuted, our entire horizon of expectations is not eliminated. On the contrary an existing system lives on in Popper's epistemology and, as we shall see in what follows, profoundly influences the fate of conjectures advanced to deal with its blemishes.

2.2 *Testing, Refutation and Confirmation*

The situation which we have now reached is the following. A problem has arisen within a widely-accepted and established theory; it has run into empirical difficulty, shall we say. To solve this problem a bold conjecture has been advanced. Its fate will ultimately be decided in a test situation of the kind shortly to be described. However, before one even goes to the trouble of obtaining empirical evidence for or against it, one can weed out hypotheses that do not meet a number of so-called 'formal' requirements of acceptability.

To appreciate the significance of these requirements, consider a situation in which a variety of conjectures is put forward in response to a problem. Then it is not necessary, or reasonable, for us immediately to test all of them. 'Even before a theory has ever undergone an empirical test,' writes Popper, 'we may be able to say whether, provided it passes certain specified tests, it would be an improvement on other theories with which we are acquainted'.[21] This is because to be considered as a reasonable alternative to the prevailing view it is, minimally, demanded by Popper of a new theory 'that it solves those problems which its predecessor solved *and* those which it failed to solve'.[22] But this is not all. There are two further so-called 'formal' requirements which it must meet: the requirement of simplicity, and the requirement of independent testability. By the first, Popper means that the new theory should unify two previously unrelated theories: the example which he likes to quote is Newton's unification of Kepler's celestial theories with Galileo's terrestrial ones. And, at its most straightforward, the requirement of independent testability amounts to a demand that new and unexpected predictions be derivable from the novel conjecture. Both of these conditions, if met, indicate that an increase in content has been achieved. Even if the conjecture in question is subsequently refuted in a test situation, it still represents a potential step forward and is worthy of serious consideration.

Popper's requirement of independent testability is formulated, at least to some extent, in the light of his profound distrust for corroborating evidence. This is because, as far as he is concerned, confirmations are all too easily come by. In particular, Popper suggests that the Marxists and the Freudians and Adlerians whom he met as a young man were able to find confirming evidence for their views wherever they looked. Whatever happened seemed to verify their theories, which appeared to be immune to revision in the light of experience. Einstein's conjecture, by contrast, exposed itself to the possibility of refutation by making risky predictions, i.e., by predicting results which were quite unexpected in the light of existing knowledge at the time. One such prediction was that light rays were attracted by heavy bodies like the Sun, a prediction which was confirmed by Eddington's eclipse experiments in 1919. Confirmations of such predictions, unlike confirmations which simply reinforced existing expectations, provide genuine empirical support for a theory.

As Popper puts it 'Confirmations should count only if they are the result of *risky predictions*; that is to say, if, unenlightened by the theory in question, we should have expected an event which was incompatible with the theory — an event which would have refuted the theory'.[23]

An example will help to bring out the significance of this claim. According to the Ptolemaic conception of the universe the Earth is stationary at the centre of a finite Universe, with Mercury, Venus, the Sun and the other planets moving in roughly concentric circles around it. Venus is deemed to be always between the Sun and the Earth, and is not expected to show phases like those of the Moon.[24] Naked eye observation bore this out: the brightness of Venus did not vary significantly. According to Copernican (and Tychonic) theory, however, Venus circled the Sun, and could be located on the opposite side of it from the Earth. Thus it was expected to show phases, which were in fact observed with the telescope. Now on Popper's view the prediction that Venus should show phases is a risky prediction — in fact, it is an example of a risky prediction which he cites himself. This is because 'unenlightened by the theory in question' i.e., the Copernican (or Tychonic) theory, we — as supporters of Ptolemaic theory — should have expected Venus not to show phases, i.e., we would have expected 'an event which was incompatible with (Copernican) theory, an event which would have refuted the (Copernican) theory'. The detection of the phases of Venus, therefore, was an unexpected (in the light of a well-established theory) piece of empirical evidence which was unearthed with the new theory. As such it counts as legitimate or genuine corroborating evidence for it.

Once a new hypothesis has met these formal requirements of acceptability, we can set about testing it. To be integrated into the accepted body of science, the conjecture must pass this test; it must meet a so-called requirement of empirical success. As Popper puts it

We have no reason to regard the new theory as better than the old theory — to believe that it is nearer to the truth — until we have derived from the new theory *new predictions* which were unobtainable from the old theory (the phases of Venus . . . .) and until we have found that these new predictions were successful.[25]

It is important to note that Popper does not demand that empirical success be achieved just as soon as a new conjecture meets the formal requirements for acceptability. All that is needed is that we 'reasonably often'[26] manage to meet the requirement of empirical success. In the interim we can entertain the new proposal, though not rationally abandon the old one, until a test can be devised to adjudicate between them. Popper's conception of scientific progress is thus quite compatible with there being a substantial time-lag between the formulation of a conjecture and its subsequent refutation and elimination, or its corroboration and incorporation into the body of science.

It emerges from the above that the requirement of independent testability uses established theory as an initially unquestioned vantage point from which to assess the suitability of a novel conjecture. More specifically it is only if the new theory makes predictions that were unexpected in the light of the old that it is worthy of further consideration. This stipulation was given a more rigorous expression in Popper's *The Logic of Scientific Discovery*.[27] From a new idea, tentatively advanced, easily testable predictions are derived with the aid of initial conditions which are 'previously accepted'. Testable predictions which are not derivable from 'current theory' and, in particular, predictions which the 'current theory contradicts' are then selected for experimental scrutiny. If the experimental analysis confirms the predictions, the theory has temporarily passed its test; it is verified, and we have no reason, as yet, to discard it. However, if the derived statements are falsified by observation and experiment, then so too is the theory from which they were deduced.

There are, then, always two theories involved in a Popperian test situation. As he likes to point out, tests in his epistemology are 'crucial' in Bacon's sense. They indicate the cross-roads between what we would expect in the light of established theory and a contradictory prediction from a new, tentatively advanced hypothesis. As Popper puts it, 'the successful new predictions which we require the new theory to produce are indentical with the crucial tests which it must pass in order to become sufficiently interesting to be accepted as an advance upon its predecessor . . . . .'.[28]

The demand that permissible corroborating evidence for a

new theory be generated in crucial tests locks novel conjectures into the established body of thought. However, Popper is not satisfied with merely formal overlap. He also insists that the scientific practitioner initially adopt a hostile attitude towards new theories. Spurred on by his fear of confirmation, Popper claims that 'Every genuine *test* of a theory is an attempt to falsify it'.[29] Now since the test is 'crucial' it is not possible for the tester to *attempt* to falsify *both* theories; this is impossible, for in falsifying one the other is confirmed. What then is the 'it' which Popper says we should seek to falsify in a crucial test situation ? It is the new hypothesis ! This is clear from Popper's remark that if a crucial (test) experiment 'does not succeed in refuting the *theory in question*, if, rather, the theory is successful with its unexpected prediction — then we say that it is corroborated by the experiment'.[30] In other words, in a test situation existing theory is *taken for granted*, and one's aim is to show that the new theory, the so-called 'theory in question', is false, thereby reinforcing one's preference for the established view.

It only remains to point out how inconsistent Popper is here. He instructs us to attempt to falsify new theories because it is easy to find confirmations when we want them. But aiming to refute a new theory in a 'crucial' test situation is tantamount to wanting to confirm one's expectations, i.e., existing theory ! Thus in implementing his demand for novel theories we violate it for prevailing orthodoxy.

2.3 *Refutation, Supersession and Piecemeal Reform*

If a novel conjecture fails to pass a crucial test it is refuted and eliminated. This is the essence of Popper's method of trial and error.

On the other hand, if the hypothesis is corroborated, the established view is refuted. But it is not eliminated. One possible future for it is that it be superseded by the new hypothesis,[31] surviving as an approximation to it. 'The predecessor of every new theory', writes Popper, 'will — from the point of view of the new theory — have the character of an approximation towards this new theory'.[32] The set of constraints which he imposes on a new conjecture effectively ensures that it is meshed into existing knowledge even while going beyond it. Well-established theory is preserved, and remains for Popper an adequate representation of a level of reality less profound than that reached by its more comprehensive successor.

Popper's conception of supersession is formulated with a view to theorizing particular historical achievements in physics as paradigmatically rational and progressive. Typically those which he selects suggest a continutiy of aim and achievement from the seventeenth century to the present

> The theories of Kepler and Galileo were unified and superseded by Newton's logically stronger and better testable theory, and similarly Fresnel's and Faraday's by Maxwell's. Newton's theory, and Maxwell's, in their turn, were unified and superseded by Einstein's.[33]

Popper often illustrates the relationship of supersession with specific instances drawn from this historical sequence. For example,[34] he compares Kepler's so-called harmonic law, which relates the periodic times in which planets complete their orbits about the Sun (T) to their mean distance from it (a), with a more comprehensive law due to Newton. According to Kepler,

$$\frac{a^3}{T^2} = \text{constant}$$

However, from Newton's theory we obtain

$$\frac{a^3}{T^2} = m_0 + m_1$$

where m_0 and m_1 are the masses of the two bodies in a binary star system. If m_0 is the mass of the Sun, then the mass of the planets (m_1) must be the same for all planets for Kepler's law to be true. But they are not. The two laws thus contradict one another; Kepler's law is false and Newton's law explains why it is so. The more comprehensive theory can not only accommodate the success of its predecessor (for planets for which m_1 is much smaller than m_0). It can also explain why Kepler's law fails, and can predict states of affairs which would falsify the harmonic law, which are unexpected in its light.

However, despite its drawbacks, Kepler's law was a magnificent achievement. Though refuted, and superseded by Newton's law, it is a good approximation to it under certain specified circumstances, and its survival in the edifice of science is guaranteed.

Consistent with the spirit if not the letter of these sentiments is Popper's commitment to piecemeal reform. He opposes this to what he calls 'holism', which he characterizes as the 'dogma of the "global" character of all tests or counterexamples'.[35] That Popper will have nothing to do with holism is apparent, in the first instance, from his discussion of the Duhem–Quine hypothesis. If a theoretical system T entails a proposition that p, and not–p, then by the logical principle of *modus tollens*, not–T: the theory is falsified. However, to achieve falsification two conditions have to be met. Firstly, it has to be accepted that the state of affairs (in empirical science) described as not–p prevails. And secondly it has to be agreed that T does, in fact, entail that p.

Quine pointed out some of the difficulties surrounding the satisfaction of these conditions. He claimed that our assertions about the external world are not tested individually and in isolation but as parts of an interconnected system of statements. Such theoretical networks, he said, impinge on reality only at the periphery, as it were. When a derived observation statement clashes with experience we may discredit the experience as, say, hallucinatory. Alternatively, we accept it, and revise some statement within the body of the system. But just what statement we should revise is not obvious — 'the total field is so underdetermined by its boundary conditions, experience, that there is much latitude of choice as to what statements to re-evaluate in the light of any single contrary experience'.[36] What that choice is to be Quine does not say, although he does imply that some piece of the system is to blame, and that a relatively minor modification rather than wholesale elimination is called for.

Like Quine, Duhem also stressed that scientists have a certain amount of freedom when dealing with theories which clash with experience. But Duhem goes beyond merely recording the fact into discussing the processes whereby decisions can be taken once it is conceded that not–p. What he emphasizes is that it can take time before the most advisable course of action emerges. In the interim a period of controversy and disagreement is likely, with different factions aligning themselves along various, and sometimes even opposing, axes. Ultimately so-called 'good sense' favours one option rather than another, although Duhem does not attempt to legislate in advance for any particular course of action. In certain circumstances, he says, a difficulty can perhaps be accommodated by making minor modifications to a system; in others this may not be good enough. A physicist, for example, who maintains obstinately and 'at any cost, at the price of continual repairs and many tangled-up stays, the worm-eaten columns of a building tottering in every part, when by razing these columns it would be possible to construct a simple, elegant, and solid system' is, says Duhem, being 'childish and unreasonable'.[37]

What Duhem and Quine emphasize, then, is that theories are vast, rambling interconnected systems of statements, and that recalcitrant observations are points at which this entire structure intersects with reality. For Duhem and for Quine this claim is logically independent of the further assertion that the only possible thing to do in such circumstances is to jettison the whole theoretical system. But it is not for Popper. For him holism means both that a theoretical system of statements faces the tribunal of sense experience as a corporate body *and* that the entire edifice must be wiped out in the face of recalcitrant evidence. And it is this conception of holism which he mistakenly attributes to Duhem and Quine. Consequently he thinks that he is criticizing them by showing that it is possible to isolate a part of a theoretical system and to hold it responsible for the falsification: a point which neither Duhem nor Quine would dispute.

Be that as it may, Popper insists both that it is *possible* merely to falsify parts of a theoretical system and that we *must* do so if we wish to proceed rationally. Taking the second point first, we find Popper writing that

> While discussing a problem we always accept (if only temporarily) all kinds of things as *unproblematic*: they constitute for the time being, and for the discussion of this particular problem what I call our *background knowledge* . . . . Almost all of the vast amount of background knowledge which we constantly use in any informal discussion will, for practical reasons, necessarily remain unquestioned; and the misguided attempt to question it

all — that is to say, *to start from scratch* — can easily lead to the breakdown of a critical debate. (Were we to start the race where Adam started, I know of no reason why we should get any further than Adam did.) . . . . Thus all criticism must be piecemeal (as against the holistic view of Duhem and of Quine); which is only another way of saying that the fundamental maxim of every critical discussion is that we should stick to our problem, and that we should subdivide it, if practicable, and try to solve no more than one problem at a time . . . .[38]

I pointed out in Section 2.1 that, in the absence of an alternative, a clash between established theory and experience generates what Popper calls a problem. Problems like this trigger the process of hypothesis formation. We now see that, for Popper, before putting forward a conjecture scientific practitioners should try to isolate the source of the problem. They should try to establish just what it is about their theory which has precipitated the difficulty, pinning down that part of it for which an alternative is sought.

Like Duhem and Quine, Popper recognizes that it is not always easy to do this. For example, he concedes that an entire theoretical system is put under strain by recalcitrant evidence, but suggests that

if we test two such systems which differ in one hypothesis only, and if we can design experiments which refute the first system while leaving the second very well corroborated, then we may be on reasonably safe ground if we attribute the failure of the first system to that hypothesis in which it differs from the other.[39]

On the other hand, if we are not fortunate enough to have two systems which differ in this way from one another, somewhat more arbitrary measures may be required. Sometimes when a system is being tested 'it is sheer guesswork which of its ingredients should be held responsible for any falsification'.[40] What's more

it is often only the scientific instinct of the investigator (influenced, of course, by the results of testing and re-testing) that makes him guess which statements . . . he should regard as innocuous, and which he should regard as being in need of modification.[41]

In other words although Popper is firmly convinced that we can and must criticize a theoretical system piecemeal, he also concedes that this can be something of a hit-and-miss affair.

From the above it emerges that, depending on the context, Popper advocates (at least) two quite different responses to the empirical failure of a well-established theory. With one eye on the history of science since the seventeenth century, he suggests that existing theory is superseded, and survives as an approximation to its richer successor. On the other hand, when he is arguing against holism, he advocates piecemeal reform of an already accepted body of thought. Either way, an established theory is not eliminated. In the struggle for survival its continued existence is secure, thought it may live on in something of a modified form. It is novel conjectures which are eliminated when they run into empirical difficulty, not tradition.

2.4 Popper's Conception of Rational Progress

Generally speaking, Popper's conception of rationality is restricted by sympathetic commentators and critics alike to what he calls the 'critical approach'. That approach is underpinned by the formal principle of *modus tollens*. It shapes Popper's claim that science progresses by the method of trial and error, of boldly proposing theories and of trying our best to falsify them. However, the discussion above reveals that there is far more involved than this when it comes to assessing whether or not someone is responding rationally in a problem-situation. By pulling together the several threads of the preceding discussion a more comprehensive picture of Popper's conception of rational progress will emerge.

At every instant of their lives socialized human beings, and scientists in particular, live at the centre of a frame of reference constituted by the existing well-established theories of the day. The periphery of the frame defines an horizon of expectations which guides scientists in their work, bestowing direction and purpose on their activity. From time to time these expectations are disappointed though — for example, when a prediction from an existing theory clashes with experience or, more precisely, with an observation statement. In the absence of an alternative, such disappointed expectations are to be treated as problems, not as refutations. They trigger the process of conjecture-formation, in which ideally several tentative theories are put forward as potential solutions to the problem in question.

To be seriously entertained a conjecture has to meet certain formal criteria. Firstly, it must preserve the successes of the existing theory, and it must also succeed where its entrenched rival failed. But more than this is required. Typically, it must also predict new effects, effects which are unexpected in the light of existing theory and which, preferably, contradict that theory. Conjectures that lock into existing theory in this way are worthy of further consideration, and they are potential steps forward. Test situations are devised in which the aim is to falsify the conjecture, to refute it. Since the new tentative theory has generated at least one prediction which contradicts prevailing orthodoxy, such tests will amount to crucial experiments between the rivals.

If the conjecture fails its test, it is rejected and eliminated, and we have no reason to abandon the old theory, even though it is still unsatisfactory and problem-ridden. However, if the conjecture is reasonably often corroborated in crucial test situations we can be sure that its empirical support is not *ad hoc*, and that it is the outcome of a genuine, independent test. Its confirmation and the refutation of existing theory go hand in hand under such circumstances. That granted, it is rational to switch allegiance to the new theory.

Though refuted, the old theory is not completely eliminated or destroyed — that would be absurd anyway, since to eliminate it would amount to eliminating those parts of the new theory which preserve its successes. Rather it is either modified in a piecemeal way, or it is superseded by its successful rival, to which it remains a good approximation in specified domains. Correlatively, the new theory is incorporated into the scientific tradition, opening up new horizons from which previously unexpected problems of even greater depth emerge. Thereupon the process just outlined is repeated. Thus we may say with Popper that (rationally) 'the growth of knowledge proceeds from old problems to new problems, by means of conjectures and refutations'.[42]

To my mind, the account just given encapsulates a conception of scientific progress which has always been present in Popper's work with varying degrees of explicitness. As here presented it fills out his tetradic schema $P_1 \rightarrow TT \rightarrow EE \rightarrow P_2$, placing it in the context of his epistemology as a whole. The analysis reveals how one-sided and misleading it is to limit Popper's view of progress to the slogan 'conjectures and refutations'. Crucially, this slogan masks the fact that, for Popper, conjecture, refutation and theory elimination are concepts delimiting the field on which novel ideas struggle for survival. These notions have to be supplemented by the triad of well-established theory, refutation and supersession, if we are to have anything like a complete picture of Popper's conception of rational scientific progress.

The importance which Popper attaches to the preservation of tradition is particularly clear in his recently developed evolutionary epistemology, or so-called 'third-world' conception of human knowledge. World 3 is, minimally, the world of cognitive entities, which Popper distinguishes from the world of mental entities (World 2), and the world of physical entities (World 1). More generally,

World 3 . . . is the world of ideas, art, science, language, ethics, institutions — the whole cutural heritage, in short — in so far as this is encoded and preserved in such World 1 objects as brains, books, machines, films, computers, pictures and records of every kind.[43]

What's more, Popper is emphatic that, although created by us, World 3 is largely autonomous. Our cultural heritage stands as an already given, taken-for-granted 'objective' reality, which constrains us. Individual contributions to its development are 'vanishingly small'. We can progressively change it bit by bit: 'we are workers who are adding to the growth of objective knowledge as masons work on a cathedral'.[44] What we cannot do is to lay the foundations for a rival structure which, as it develops, will entail the ultimate ruin of the existing edifice.

Feyerabend has recently claimed that modern scientists are narrowly sceptical, directing their doubts towards new ideas, and tolerating minor modifications of their own basic ideas, but never questioning those basic ideas themselves.[45] If that is true, and if my interpretation of Popper is correct, they are not that dissimilar to paragons of Popperian rationality. For, on the one hand, Popper's conception of rational behaviour strongly encourages an even hostile scepticism to novelty. Hostile, because rather than requiring that tests of conjectures should be 'open-minded' attempts to establish whether they are false *or true*, Popper repeatedly emphasizes that 'genuine' tests of new theories are attempts to refute them with a view to eliminating them. And, on the other hand, Popper holds that the rational advance of knowledge 'consists, mainly, in the modification of earlier knowledge'. Admittedly in Popper's case this can involve more than simply 'minor modifications' of one's basic ideas. But the effects of revolutionary changes at this deep level are not particularly traumatic, since a well-entrenched theory is reformed, or superseded by its successful new rivals; it is not replaced. 'The rational procedure', as Popper puts it, 'is to correct and to revolutionize it, but not to sweep it away'.

2.5 Popper's Continuism

I pointed out earlier that a key feature of revolutionary change is that it involves the rejection of one system and its replacement by another. Popper sees this. What's more, in line with the 'liberal' conception of revolution he tends to suggest that the entire process occurs very rapidly. Hence Lakatos' remark, quoted earlier, that for Popper progress occurs by the 'incessant' production of new ideas and their 'fast elimination' if they are empirically defective.

Now if Popper were to advocate that existing theories be revolutionized in this sense, it would amount to his demanding that they be rejected and replaced in one fell swoop. But he will have none of that, and rightly so. Popper is emphatic that it would be fatal to wipe out all that we know suddenly. And since he (mistakenly, to my mind) tends to identify a change of that kind with revolutionary change, for him a traditional body of thought is never eliminated and replaced. It is reformed or superseded.

The conception of rational progress which emerges from (my rational reconstruction of) Popper's work is thus an essentially continuist one. The view of historical change that it embodies is basically that which Randall attributes to the development of science from Greek times to the present. Scientific advance, he writes, involves the 'co-operative criticism and reconstruction of a well-organized system of ideas, shaken from time to time by fresh insights which have had to be worked into the logical structure'.[46] Where Randall speaks of fresh insights Popper speaks of bold conjectures springing from the imagination, or from poetic intuition. It is they that shake the existing edifice of science. But they do not topple it. For no sooner is the new idea put forward, than it must come under the sway of established tradition. As the bearers of that tradition, the members of the scientific community critically evaluate the revolutionary new proposal. Reason demands that it be rejected and eliminated, or that it be worked into the logical structure of the existing system of thought. According to Popper, it is only by thus incorporating new insights into the framework of well-established theories that the rational — and continuous — progress of science will be ensured. His 'revolutions' are restricted to the peripheral terrain on which conjectures compete with one another for survival. They are, as Foucault has put it, little more than 'moments of consciousness'.[47]

From the vantage point adopted here, Popper's conception of what constitutes rational progress in scientific life converges with his view of rational social change. Like their scientific counterparts, revolutionaries in society who wish to create a social *tabula rasa*, he says

> should not be surprised if they find that once they destroy tradition, civilisation disappears with it. They will find that mankind have returned to the position in which Adam and Eve began — or, using less biblical language, that they have returned to the beasts. All that these revolutionary progressivists will then be able to do is to begin the slow progress of human evolution again.[48]

The social institutions that we have must be changed gradually, and the consequences of our minor reforms carefully monitored. All changes in society have unexpected and unwanted consequences, and it is these that we must be on the look-out for. To enable us to identify just what it is that we have done that has precipitated them, 'piecemeal social engineering' is imperative. If we are more ambitious, if we are 'holistic' and Utopian, we will rapidly lose control over the pace and direction of social change. To achieve our objectives reasoning will then give way to violence and terror as we struggle to impose our grand blueprint on a recalcitrant populace.

I pointed out above that Popper recognizes that it is not easy to localize the source of a defect in a scientific theory. His appeal to instinct and guesswork in that context is reflected in his reliance on 'piecemeal tinkering' and the 'spirit of "muddling through"'[49] in the social field. As Carr puts it

> The status of reason in Professor Popper's scheme of things is, in fact, rather like that of a British civil servant, qualified to administer the policies of the government in power and even to suggest practical improvements to make them work better, but not to question their fundamental presuppositions or ultimate purposes . . . [However] progress in human affairs, whether in science or in history or in society, has come mainly through the bold readiness of human beings not to confine themselves to seeking piecemeal improvements in the way things are done, but to present fundamental challenges in the name of reason to the current way of doing things and to the avowed or hidden assumptions on which it rests.[50]

Looking back over the colossal changes in science in which he had participated, Werner Heisenberg remarked that

> Only those revolutions in science will prove fruitful and beneficial whose investigators try to change as little as possible and limit themselves to the solution of a particular and clearly defined problem. Any attempt to make a clean sweep of everything or to change things arbitrarily leads to utter confusion.[51]

In doing so, suggests Feuer, he was advocating what is called among political scientists 'the doctrine of conservative revolution' — the view that the only fruitful revolutions are those that are carried out with a minimum of revolutionary intent and with a minimum disruption of the *status quo*. Popper, too, is a so-called conservative revolutionary.

2.6 Conclusion

As I pointed out above, Popper's retreat to continuism is a direct consequence of his commitment to the 'liberal' conception of revolution. If invoked to theorize the changes affecting a traditional body of thought, this conception implies that all that has been painstakingly built up during previous periods of relative tranquility is destroyed in one fell swoop. Popper rightly refuses to incorporate that idea into a conception of rational progress. Yet he believes that revolutions have an important role to play in the growth of knowledge. To reconcile these demands he is forced to restrict revolutions (taken in the 'liberal' sense) to the peripheral yet important battles between novel conjectures. Accordingly his theory of rational growth does not allow for nor permit historical discontinuity.

If the only alternative to preserving an established theory was to wipe it away overnight we could hardly blame Popper or anyone else for opposing discontinuous transitions. But as I stressed in Chapter 1, it is not the only alternative. For it is possible to reconcile discontinuity with 'gradualism', with a conception of change in which the established order, though not rapidly destroyed, is nevertheless eliminated root and branch, and replaced. This process takes time; the old withers away as the new gradually establishes its superiority.

It is to Popper's credit that he developed a conception of rational progress which emphasized the importance of critical opposition and theory elimination, while at the same time insisting that one should not aim to build an entire new order from scratch. His failure nevertheless to transcend continuism is due to the pressures exerted by a defunct conception of revolutionary change, and an associated conception of reason, as discussed in Chapter 1. Neither Popper nor Lakatos, to whom I now turn, has escaped their grip.

Notes

1. K.R. Popper, *Conjectures and Refutations* (Routledge & Kegan Paul, London, fifth edition, 1974), p.215.
2. *Ibid*. p.129.
3. *Ibid*. p.238.
4. K.R. Popper, *Objective Knowledge* (Clarendon Press, Oxford, 1972), p.345.
5. K.R. Popper, *Conjectures, op.cit.* p.313.
6. *Ibid*. p.141.
7. I. Lakatos, 'Popper on demarcation and induction', in P.A. Schilpp (ed), *The Philosophy of Karl Popper* (Open Court, Illinois, 1974), pp.242–3.
8. I. Lakatos, 'Falsification and the methodology of scientific research programmes', in I. Lakatos and A. Musgrave, (eds) *Criticism and the Growth of Knowledge* (Cambridge University Press, Cambridge, 1973), pp.173 *et seq*.
9. I. Lakatos, 'Popper', *op.cit.* p.245.
10. K.R. Popper, 'The Rationality of Scientific Revolutions', in R Harré (ed), *Problems of Scientific Revolution* (Oxford, Clarendon Press, 1975), p.83.

11   Ibid. p.93.
12   K.R. Popper, *Conjectures*, op.cit. p.32.
13   Ibid. p.28.
14   See, for example, K.R. Popper, *Objective Knowledge*, op.cit. pp.164, 243, 287.
15   Ibid. p.243, for example.
16   K.R. Popper, *Conjectures*, op.cit. pp. vii, 313.
17   Ibid. p.132.
18   Ibid. p.222.
19   K.R. Popper, *Objective Knowledge*, op.cit. pp.344, 345.
20   Ibid. pp.346–7.
21   K.R. Popper, *Conjectures*, op.cit. p.217.
22   K.R. Popper, *Objective Knowledge*, op.cit. p.15.
23   K.R. Popper, *Conjectures*, op.cit. p.36.
24   T.S. Kuhn, *The Copernican Revolution* (Harvard University Press, Cambridge, Mass., 1957), p.223.
25   K.R. Popper, *Conjectures*, op.cit. p.246.
26   Ibid. p.243.
27   K.R. Popper, *The Logic of Scientific Discovery* (Hutchinson, London, 1959), paragraph 3.
28   K.R. Popper, *Conjectures*, op. cit. p.247.
29   Ibid. p.36.
30   Ibid. p.112, my italics.
31   Popper actually lists a large number of conditions for supersession in his *Conjectures*, p.232. However, the formal and material requirements which I have stated seem to be the most important ones as far as he is concerned.
32   K.R. Popper, *Objective Knowledge*, op.cit. pp.16–7.
33   K.R. Popper, *Conjectures*, op.cit. p.220.
34   K.R. Popper, *Objective Knowledge*, op.cit. pp.200–2.
35   K.R. Popper, *Conjectures*, op.cit. p.239.
36   W.V.O. Quine, *From a Logical Point of View* (Harper & Row, New York, 1961), pp.42–3.
37   P. Duhem, *The Aim and Structure of Physical Theory* (Princeton University Press, Princeton, 1954), p.217.
38   K.R. Popper, *Conjectures*, p.238.
39   K.R. Popper, *The Poverty of Historicism* (Routledge & Kegan Paul, London, second edition, 1960), p.132.
40   K.R. Popper, *Conjectures*, op.cit. p.239.
41   K.R. Popper, *The Logic*, op.cit. p.76.
42   K.R. Popper, *Objective Knowledge*, op.cit. p.258, original emphasis suppressed.
43   B. Magee, *Popper* (Fontana, London, 1973), p.61.
44   K.R. Popper, *Objective Knowledge*, op.cit. p.121.
45   P. Feyerabend, *Against Method* (New Left Books, London, 1975), p.298.
46   J.H. Randall, Jr, 'The development of scientific thought in the School of Padua', *Journal of the History of Ideas*, 1 (1940), 179.
47   M. Foucault, *The Archaeology of Knowledge* (Tavistock Publications,

London, 1974), p.12.
48 K.R. Popper, *Conjectures*, *op.cit.* p.344.
49 K.R. Popper, *The Poverty*, especially Chapter 3.
50 E.H. Carr, *What is History* ?, (Penguin Books, Harmondsworth, 1964), p.155.
51 W. Heisenberg, quoted in L.S. Feuer, *Einstein and the Generations of Science* (Basic Books Inc, New York, 1974), p.295.

3 LAKATOS' QUEST FOR CONTINUITY

LAKATOS' theory of scientific progress represents one of the most sustained attempts to defend the rationality of science as we know it. It is constructed, in the main, by critically assimilating and transforming the views of Popper, Kuhn and Feyerabend in the light of ('rationally reconstructed') episodes in the history of science. Lakatos is apparently at home in the philosophy both of mathematics and of the empirical sciences. The historical case studies that he develops are drawn from physics and from chemistry, and he unhesitatingly discusses in some detail the Scientific Revolution as well as developments in physics in this century. Lakatos' style is polemical and aggressive if somewhat rambling. By all accounts he was an exhilirating and entertaining lecturer and friend and in an obituary written after his unexpected death in 1974, Feyerabend described him as 'an outstanding thinker and the best philosopher of science of our strange and uncomfortable century'.[1] And, predictably, an immense volume of articles dedicated to Imre Lakatos has already appeared.

It is difficult not to be dazzled by Lakatos' achievements, and objectively to evaluate his contribution to the current debate in the philosophy of science. However, as the argument in this chapter unfolds it will, I think, become clear that the importance of his work has been grossly overrated by his more enthusiastic admirers.

Lakatos' overall aim is to show that 'rationality works much slower than most people tend to think, and, even then, fallibly'.[2] That granted, he is confronted with several options as far as revolutionary change is concerned. One thing he could do is to follow the approach I advocated in Chapter 1. In other words he could accept that there are discontinuous transitions in science, and that they are rationally intelligible. This would involve his then developing a 'non-liberal' conception of revolution embodying the idea — consistent with Lakatos' view of rational change — that it takes time to eliminate one system of relationships and to replace it with another.

Lakatos does not do this, however. Rather he remains effectively wedded to the 'liberal' conception of revolution. From this perspective, as I have pointed out, elimination and replacement occurs fairly rapidly. Given his view that reason works slowly, this presents Lakatos with something of a dilemma. Crudely put, he has either to deny that revolutions are rationally intelligible or, to preserve the rationality of scientific change as he sees it, he has to deny that revolutions occur at all.

In this chapter it is argued that Lakatos oscillates somewhat between these positions, ultimately drifting, perhaps only half-consciously, into the latter alternative. To theorize scientific change he develops his so-called methodology of scientific research programmes to account for continuity in science and the tenacity of some scientific theories. This methodology, as we shall see, is based on Popper's work and, in its original formulation, is intended to demarcate progressive from degenerating (Kuhnian) normal science. However, Lakatos rather haphazardly transfers it out of this context into that of revolutionary change. With that his difficulties begin. Having laid such stress on continuity, he finds himself, on the one hand, in the embarrassing position of being unable rationally to eliminate or supersede one system of thought (or research programme, as he calls it) by another. This forces him far closer to Kuhn than he would care to admit. And, on the other hand, in collaboration with Zahar, he formulates a conception of rational progress that effectively dissolves revolutionary transformations altogether. The whole of science begins to take on the appearance of Kuhnian 'normal' science. In the end, Lakatos' quest for continuity becomes something of an obsession which destroys the coherence of his project.

3.1 *The Critique of Popper*

The specific cluster of ideas which Lakatos finds objectionable in Popper's work are those centering on the latter's claim that, to facilitate progress in science

> there is no more rational procedure than the method of trial and error — of conjecture and refutation: of boldly proposing theories; of trying our best to show that these are erroneous; and of accepting them tentatively if our

critical efforts are unsuccessful . . . . If the outcome of a test shows that the theory is erroneous, then it is eliminated; the method of trial and error is essentially a method of elimination.[3]

This formulation of his position by Popper is misleading. It abstracts that part of his methodology which deals with the fate of novel conjectures from his overall conception of rational progress (see Chapter 2). Kuhn does not appreciate this, as we shall see. His mistake is understandable though, given the unsystematic way in which Popper sets out his views.

One notable feature of the slogan 'conjecture, refutation, theory elimination' is that it locates the blame for a clash between theory and experiment squarely on the shoulders of the theory.[4] As such it preserves what Popper calls the principle of empiricism. According to this principle, 'the fate of a theory, its acceptance or rejection, is decided by observation and experiment — by the result of tests'.[5] If a (conjectural) theory fails to pass its test, it is eliminated. For Popper, then, observation and experiment play a dominant role in the revolutionary process.

Lakatos' critique of these views of Popper's has been strongly influenced by Kuhn's opposition to them. Kuhn pointed out that his empirical investigations had revealed that

No process yet disclosed by the historical study of scientific development at all resembles the methodological stereotype of falsification by direct comparison with nature . . . . The decision to reject one paradigm is always simultaneously the decision to accept another, and the judgment leading to that decision involves the comparison of both paradigms with nature *and* with each other.[6]

Kuhn does not realize that this conception of scientific growth is quite compatible with Popper's views or, rather, with his views as 'rationally reconstructed' in Chapter 2. Where he differs significantly from Popper is in his suggestion that the 'decision', if one can call it that, to switch from one paradigm to another is not steered by arguments based on logic and experience alone, nor should it necessarily be. As I pointed out in Chapter 1, Kuhn believes that, in the last analysis, the rejection of one conceptual framework and its replacement by another involves a non-rational process

LAKATOS' QUEST FOR CONTINUITY

something like a gestalt switch. This means that for him scientific revolutions, which are central episodes in the advance of knowledge, ultimately lie beyond the bounds of rational intelligibility.

Those who believe in the essential rationality of science and of its history, like Lakatos, were particularly disturbed by Kuhn's argument. Rightly or wrongly, Lakatos felt that Kuhn had reduced major scientific change to a matter of mob psychology, and had also inadvertently provided ammunition for student revolutionaries, whom Lakatos called 'contemporary religious maniacs'. This stimulated him to look more closely at Popper's epistemology in an attempt to develop a theory of rational scientific progress.

Lakatos conceded Kuhn's point against Popper: yes, he argued there is a 'naïve' strand in Popper's theory of progress which decrees that development is to be achieved solely by conjecture, refutation and theory elimination in the light of recalcitrant evidence.[7] In its terms advances in science involve a confrontation between a single theory and experiment, and the only worthwhile outcome of such a confrontation is the refutation and elimination of the theory. Lakatos conceded that *if* we restrict the scope of rational response in this way then, with Kuhn, we cannot but regard major scientific advances as irrational. The history of science shows that neither of the requirements of naïve falsificationism are satisfied in practice. In the first place, tests are found usually to involve a struggle between at least two rival theories and experiment: as Kuhn put it in the earlier quotation 'The decision to reject one paradigm is always simultaneously the decision to accept another, and the judgement leading to that decision involves the comparison of both paradigms with nature *and* with each other'. And secondly, as a corollary of this, some of the most important outcomes of such tests involve confirmation rather than refutation. In other words, if Popper's 'naïve' conception of rational progress wers correct, then scientific advance would indeed be irrational, as Kuhn claimed.

As far as Lakatos is concerned, then, Kuhn's emphasis on the irrational in history reflects the narrowness of Popper's 'naïve' conception of rationality. By broadening the conception of rational progress, we can restrict the scope of, though never fully eliminate, the irrational component in the growth of scientific knowledge. Some of the material for that broader concept of rationality is contained in the work of Popper himself. Kuhn, writes Lakatos, 'is right in objecting to naïve falsificationism, and also in stressing the *continuity* of scientific growth, the *tenacity* of some scientific theories. But Kuhn is wrong in thinking that by discarding naïve falsificationism he has discarded thereby all brands of falsificationism'.[8] For within Popper's work there is another brand of falsificationism, the 'sophisticated' variant, which Popper 'conflated' with the naïve version. According to Lakatos 'sophisticated methodological falsificationism' provides the basis for the rational reconstruction of the history of science, combining Kuhnian demands for continuity and tenacity with a set of criteria for rationally appraising theory change.

Lakatos' critique of Kuhn reflects his awareness of that dimension of Popper's thought on which I placed such emphasis in Chapter 2. As I pointed out there, Popper is not a 'naïve' falsificationist, although he often presents his views as if he were. And his conception of rational progress is one which, in fact, stresses continuity, and advocates the tenacious preservation in some form of a deeply entrenched system of thought. In struggling to extract this aspect of his mentor's epistemology from Popper's scattered comments, Lakatos formulates his own methodology. Unlike Popper, he applies it indifferently to established theories and to novel conjectures, as we shall see. The result is predictable. Popper's revolutionary rhetoric (of rejection and replacement) was confined to the terrain on which new ideas struggle for survival. By importing continuity and tenacity into that domain as well, Lakatos effectively exorcizes revolutions from science altogether.

3.2 Lakatos' Methodology: The Articulation of a Paradigm

The tendency to see Popper as an advocate of 'naïve' falsificationism lies in his emphasis on the importance to science of conjecture, refutation and theory elimination. In *The Logic of Scientific Discovery*, Popper laid particular stress on this dimension of his methodology. It was there that he first invoked the logical principle of *modus tollens*, and proposed falsifiability as a demarcation criterion between scientific and other systems of thought.

Popper's demarcation criterion has a double aspect. In part, it is 'formal', in that it is underpinned by *modus tollens*. However, Popper is also acutely aware that 'no conclusive disproof of a theory can every be produced'[9] — that the implementation of this logical principle can be thwarted by ingenious and imaginative scientists. And so he formulated methodological rules to back up his formal considerations. These ban strategies for evading falsification and theory elimination. If these rules are adhered to, the fate of theories will indeed be dictated by observation and experiment in conjunction with the canons of logic. Scientists will not allow themselves to engage in wily manoeuvres to spare refuted conjectures. They will ruthlessly discard them in the interest of rational progress. In doing so they evince the critical approach which, says Popper, is the 'proper one for the scientist'.

The seeds of sophisticated falsificationism also lie in Popper's earliest work. For in *The Logic of Scientific Discovery* he did, as a matter of fact, permit responses to contrary evidence which did not involve falsification and ruthless theory elimination. These arose in the context of his discussion of conventionalism. According to Popper, the conventionalist regards scientific theories as systems of implicit definitions or of analytic statements. A theory interpreted or re-interpreted in this way would be unfalsifiable by empirical methods (and thus non-scientific on Popper's demarcation criterion between science and metaphysics). Conventionalists thus have at their disposal a way of saving a theory from refutation when it clashes with experience. Popper claims that they will be particularly prone to invoke it, and any number of other 'immunizing strategies' or 'conventionalist twists' when an established theory is jeopardized by recalcitrant experience

Whenever the 'classical' system of the day is threatened by the results of new experiments which might be interpreted as falsifications according to my point of view, the system will appear unshaken to the conventionalist. He will explain away the inconsistencies which may have arisen; perhaps by blaming our inadequate mastery of the system. Or he will eliminate them by suggesting *ad hoc* the adoption of certain auxiliary hypotheses, or perhaps of certain corrections to our measuring instruments.[10]

Regarding the first and third of these manouevres to evade falsification, Popper was uncompromising, insisting that 'Intersubjectively testable experiments are either to be accepted, or to be rejected in the light of counter-experiments. The bare appeal to logical derivations to be discovered in the future can be disregarded'.[11] But he was more lenient when it came to the question of auxiliary hypotheses; these could be introduced provided they were not *ad hoc*. The corresponding methodological rule reads

> As regards *auxiliary hypotheses* we propose to lay down the rule that only those are acceptable whose introduction does not diminish the degree of falsifiability or testability of the system in question, but, on the contrary, increases it . . . The introduction of an auxiliary hypothesis should always be regarded as an attempt to construct a new system; and this new system should then always be judged on the issue of whether it would, if adopted, constitute a real advance in our knowledge of the world.[12]

In similar vein, Popper allowed changes in the '*explicit* definitions whereby the concepts of an axiom system are given a meaning in terms of a system of lower level universality. Changes in these definitions are permissible if useful; but they must be regarded as modifications of the system, which thereafter has to be re-examined as if it were new'.[13]

The position which Popper adopts here on auxiliary hypotheses and explicit definitions is significant. When he first introduced the idea of methodological rules in *The Logic of Scientific Discovery*, it was in a context in which he condemned Newtonians for defending classical physics against the Einsteinian alternative. By attempting to sidestep recalcitrant evidence, he said, they had elevated an 'obsolete scientific theory' into a 'form of metaphysics'.[14] This context, and Popper's general interest in scientific revolutions, readily leads one to think that methodological rules are primarily intended to ensure that entire systems of thought are falsified and eliminated (or superseded) when they clash with experience. But as we can see from the above, this is not necessarily so: *modifications*, and minor ones at that, are also permitted on condition that they are not *ad hoc*. And to settle the question of *ad hoc*-ness it is necessary to compare the system as modified with its predecessor, to see whether or not progress (as measured in *The Logic of Scientific Discovery* by the degree of falsifiability) has been made.

LAKATOS' QUEST FOR CONTINUITY

This context provides the basis for what Lakatos calls 'sophisticated methodological falsificationism'. He notes that falsification is not the only possible response to a clash with experience: 'Why not', he writes, 'rather impose certain standards on the theoretical adjustments by which one is allowed to save a theory ?'.[15] Referring to the sections of Popper's work which I have just discussed, Lakatos then spells out what he means by sophisticated falsificationism

> Let us take a series of theories, $T_1, T_2, T_3 \ldots$ where each subsequent theory results from adding auxiliary clauses to (or from semantical re-interpretations of) the previous theory in order to accommodate some anomaly, each theory having at least as much content as the unrefuted content of its predecessor. Let us say that such a series of theories is *theoretically progressive (or 'constitutes a theoretically progressive problemshift')* if each new theory has some excess empirical content over its predecessor, that is, if it predicts some novel, hitherto unexpected fact. Let us say that a theoretically progressive series of theories is also *empirically progressive (or 'constitutes an empirically progressive problemshift')* if some of this excess empirical content is also corroborated, that is, if each new theory leads us to the actual discovery of some *new fact*. Finally, let us call a problemshift *progressive* if it is both theoretically and empirically progressive, and *degenerating* if it is not.[16]

What has happened here is this. From the early Popper, Lakatos has derived the idea of modifying a theory when it clashes with experience by introducing auxiliary hypotheses or by redefining some of its basic terms. Popper said that after each such modification the theory should be treated as if it were a new system: hence Lakatos speaks, somewhat misleadingly, of a *series* of theories, rather than of modifications to a single theory. Popper also insisted that such modifications were only permitted if they were not *ad hoc*, and led to a real advance in our knowledge. Bearing in mind the fact that the early Popper was loath to specify criteria for rational scientific progress, defined as approaching closer to the truth, Lakatos has extracted the conditions under which adjustments are allowed from the Popper of *Conjectures and Refutations*. These were discussed at length in the previous chapter. I showed that for a new theory to be accepted into the body of a science Popper stipulates that it retain the unrefuted content of its predecessor, that it meet the 'formal' requirement of independent testability i.e. that it leads to new and unexpected predictions, and that these predictions be experimentally corroborated (the requirement of empirical success). All that Lakatos has done in the above quotation is, basically, to weld these two aspects of Popper's work together.

It is a strange alliance. The early Popper saves an established theory by allowing non- *ad hoc* modifications to it. The later Popper does so by describing clashes with experience as precipitating a problem. Problems trigger new hypotheses, and Popper lays down criteria for their acceptability in *Conjectures and Refutations*. Lakatos invokes these criteria for the rational acceptability of conjectures to assess the legitimacy of modications to an established system of thought as tolerated by the early Popper. The gulf is bridged by calling modified versions of an entrenched view 'new' theories. By doing so Lakatos is able to use the evaluative apparatus generated for new ideas to assess the rationality of relatively minor changes to old ones.

The most striking difference between naïve and sophisticated falsificationism is that whereas the former embodies a conception of rational progress which concentrates exclusively on theory *elimination*, sophisticated falsificationism provides a set of criteria by virtue of which it is rational to *save* a theory. With theory preservation as his aim, Lakatos developed his own methodology of scientific research programmes. From sophisticated falsificationism he took the criteria for progressive and degenerating problemshifts, and he integrated them into a slightly different context. Lakatos spells out the considerations which informed this change in emphasis in the following quotation

> One of the crucial features of sophisticated falsificationism is that it replaces the concept of *theory* as the basic concept of the logic of discovery by the concept of *series of theories*. *It is a succession of theories and not one given theory which is appraised as scientific or pseudo-scientific*. But the members of such series of theories are usually connected by a remarkable *continuity* which welds them into *research programmes*. This *continuity* — reminiscent of Kuhnian 'normal science' — plays a vital role in the history of science; the main problems of the logic of discovery cannot be satisfactorily discussed except in the framework of a *methodology of research programmes*.[17]

Lakatos' research programmes consist essentially of a hard core of basic axioms around which a structure is built up over time by extension and accretion. The hard core is protected from crumbling too easily by what he calls a 'negative heuristic'. This forbids scientists to revize it in the light of recalcitrant evidence; anomalies are absorbed by a protective belt of auxiliary hypotheses which can be modified when the programme runs into empirical difficulty. However, not all anomalies need be taken seriously. In fact, Lakatos insists that research programmes develop in an ocean of anomalies; it is what he calls the 'positive heuristic' which enables scientists not to become confused by this situation. It provides them with hints and suggestions on how to modify or to elaborate the protective belt surrounding the hard core. And if it is particularly powerful it can even help a programme to forge ahead more or less independently of empirical support, 'digesting' some anomalies which are thereby turned into corroborations, and shelving, or simply ignoring, others.

If the argument in Chapter 2 is correct, it emerges that Lakatos shares with the later Popper the desire to preserve in some form an established framework of ideas. Where sophisticated falsificationism really does differ from Popperian methodology is in Lakatos' willingness to shelve anomalies or to ignore them (see Kuhn). He allows the positive heuristic to develop more or less independently of empirical considerations for substantial periods of time. He even goes so far as to say 'If a scientist (or mathematician) has a positive heuristic, he refuses to be drawn into observation. He will "lie down on his couch, shut his eyes and forget about the data"'.[18]

A number of related considerations inform Lakatos' tolerant approach to recalcitrant evidence. In the first place, he is emphatic that we should treat new, budding research programmes leniently. Feyerabend has explained why

> When a theory enters the scene, it is usually somewhat inarticulate, it contains contradictions, the relation to the facts is unclear, ambiguities abound, the theory is full of faults. However it can be developed, and it may improve. The natural unit of methodological appraisals is therefore not a single theory, but a succession of theories, or a *research programme* . . .[19]

If we applied the rigid canons of naïve falsificationism to each new theory as it was proposed it would stand no chance of surviving, and we would never find out what its strengths were.

A second consideration informing Lakatos' attitude to anomaly is his desire to account for what he calls the relative autonomy of theoretical science. This is reflected in his view that the positive heuristic can steer a research programme in specific directions quite independently of observational data. That again is part of Lakatos' unwillingness unduly to restrict 'the freedom of the scientist, the freedom of daring intellectual enterprises included'.[20] Consequently, all that he requires is that

> each step of a research programme be consistently content-increasing: that each step constitute a *consistently progressive theoretical problemshift*. All we need in addition to this is that at least every now and then the increase in content should be seen to be retrospectively corroborated: the programme as a whole should also display an *intermittently progressive empirical shift*. We do not demand that each step produce *immediately* an *observed* new fact.[21]

Although the methodology of scientific research programmes stresses confirmation and not refutation, then, empirical success is subservient to theoretical innovation, which leads the way.

Finally, Lakatos' relative diffidence about recalcitrant experience reflects his allegiance to the 'requirement of continuous growth', as he calls it. The rational naïve falsificationist repeatedly breaks the rhythm of scientific advance: for him or her 'progress consists of an incessant, ruthless, revolutionary confrontation of bold, speculative theories and repeatable observations, and of the subsequent fast elimination of the defeated theories'.[22]

Lakatos is diametrically opposed to such theories of instant rationality. We can only account for the continuity of science and the tenacity of scientific theories — both of which Kuhn has brought to our attention — with the methodology of scientific research programmes. As Lakatos repeatedly points out, this methodology accommodates those stable periods which Kuhn has called normal science. But it does so within a normative framework, setting up criteria for distinguishing, says Lakatos, 'good, progressive normal science', from 'bad, degenerating normal science'.[23]

This point is important. The variant of the methodology of

scientific research programmes discussed thus far *postulates* a hard core and a positive heuristic, and adopts standards from sophisticated falsificationism for evaluating particular moves within a research programme. However, sophisticated falsificationism posits a series of theories the members of which are generated by 'adding auxiliary clauses to (or from semantical reinterpretations of) the previous theory in order to accommodate some anomaly' (see Ref. 16). It is within this context that Lakatos first introduces the concepts of progressive and degenerating problemshifts, and the notion of novel facts. The moves made within the research programme can be evaluated using this terminology, then, without any reference to a rival programme with its own hard core being implied. At this level, Lakatos' methodology effectively lays down guidelines for scientists who are, in Kuhnian language, articulating a paradigm. It is primarily concerned to assess the legitimacy of modifications within a research programme *with reference to an earlier stage of its own development*, and not to question the viability of the entire programme with its hard core. As such it has no obvious bearing on scientific revolutions. During such periods the hard core is no longer held to be irrefutable. It crumbles, to be rejected and replaced by a rival. Many of the features of Kuhn's work which Lakatos finds objectionable pertained to the former's attempt to theorize these discontinuous transitions, which Kuhn thought were ultimately not rationally intelligible. In the formulation of his methodology presented above Lakatos has not yet come to grips with this possibility; he has theorized normal science, not revolutionary science. He has imposed 'certain standards on the theoretical adjustments by which one is allowed to save a theory' (see Ref. 15); he has not yet provided us with guidelines for removing research programmes with their hard cores.

3.3 Lakatos' Methodology: Scientific Revolutions — Introductory Comments

For the present purposes scientific revolutions are (possibly protracted) periods during which two or more research programmes, with different hard cores (and, possibly, positive heuristics) compete with one another for the allegiance of the scientific community. Feyerabend has formulated a conception of Lakatos' methodology appropriate to such situations, which he presents as if it were the whole story

> The methodology of research programmes develops standards for the evaluation of (scientific, or, more generally, conceptual) *change*. The standards apply to research programmes, not to individual theories; they judge the evolution of a programme over a period of time, not its shape at a particular time; and they judge this evolution in comparison with the evolution of rivals, not by itself. A research programme is called 'progressive', if it makes predictions that are confirmed by subsequent research and thus lead to the discovery of novel facts. It is called 'stagnating' if it makes no such predictions, but is reduced to absorbing material that was discovered with the help of its rivals. Thus the first step in the application of the methodology of research programmes is to present the research programmes which are to be compared.[24]

There are some striking differences between this statement and the one in which Lakatos originally spelt out his criteria for progressive and degenerating problemshifts (Ref. 16, Section 3.2). Notably, whereas Lakatos assessed the modifications to a research programme by comparing the revized version with a previous stage in the development of the *selfsame* programme, Feyerabend does so by appealing to a *rival* programme. So, while Lakatos (in the earlier quotation) makes 'vertical' judgements which presuppose a shared hard core, Feyerabend suggests that the methodology of scientific research programmes makes 'horizontal' judgements between programmes with different hard cores. Not that Feyerabend is always consistent on this matter. Thus in one and the same paper he says that in Lakatos' methodology 'we do not judge the *state* in which a research programme finds itself at a particular moment, we judge its *history*, preferably (sic) in comparison with the history of rival programmes',[25] and that 'the methodology of research programmes does not examine research programmes by themselves, it examines them in comparison with other research programmes'.[26]

Feyerabend can hardly be blamed for this ambiguity. It simply reflects the slipshod way in which Lakatos characterizes novelty. Typically in a recent paper with Zahar on the Copernican achievement he claims that Newton's programme was progressive because 'it successfully anticipated novel facts like the return of Halley's comet, the existence and the course of Neptune and the bulge of the Earth'.[27] No reference to a rival programme is made; mere temporal novelty seems to be a sufficient condition for progress. But later in the same paper Lakatos — quite mistakenly, to my mind — claims that 'I originally defined a prediction as "novel", "stunning", or "dramatic" . . . if the predicted fact was forbidden by the rival programme'.[28] *The* rival programme ? Lakatos' standards of progress and degeneration do not presuppose a rival programme as this quotation suggests, although they have been carried over into that context. Correlatively, there is a concept of novelty in his methodology that can help us to distinguish between 'good, progressive normal science' and 'bad, degenerating normal science'. A rival programme with its own hard core is simply not necessary for this purpose.

These are not mere debating points. A conception of novelty which assesses the legitimacy of adjustments to the protective belt of a research programme by comparing the 'new' theory in the series with its predecessor, presupposes that the theories being compared share a hard core. The hard core is thus not at issue when judgements of progress are made nor, correlatively, is it implied that to achieve progress it may be necessary to reject an entire research programme with its hard core. But if novelty is characterized with reference to a rival programme, then it is the entire programme that is being judged as progressive or degenerating and, accordingly, it is implied that to achieve progress it may be necessary to jettison that programme and its hard core. From the point of view of the former conception of novelty, rational change is relatively smooth and continuous; from the point of view of the latter it is discontinuous and disruptive.

From the above it is clear that Lakatos has (at least) two different conceptions of 'novelty', each informing a different criterion of progress. He is not particularly self-conscious about this, nor does he consistently differentiate between them. One of the two is that which applies to Kuhnian normal science, when the question of eliminating a research programme does not arise. The other is that appropriate to circumstances in which there are rival programmes involved – which, according to Lakatos, is usually the case. Under these circumstances, a modification to a programme is judged as progressive or degenerating in the light of a rival. For the time being I am going to work on the assumption that Lakatos assesses progress within a programme with reference to a rival. This implies that the conception of novelty involved in such judgements is similarly underpinned. Later I shall show how Lakatos surreptitiously shelves the rival even when discussing revolutionary changes. The immediate effect of this is to move closer to the 'normal science' conception of novelty — from which perspective the hard core readily disappears from view. The sought-for continuity of history is thus achieved — but only by a sleight of hand. And with that the coherence of Lakatos' approach is lost — because of a passion for smooth and gradual change.

3.4 The Demand for Proliferation

'The first step in the application of the methodology of research programmes', writes Feyerabend, 'is to present the research programmes which are to be compared'. Quite so. As Lakatos has emphasized

Originally people had hoped that a 'logic of discovery' would provide them with a mechanical book of rules for solving problems. This hope was given up: for Popper the logic of discovery or 'methodology' consists merely of a set of (tentative and far from mechanical) rules for the appraisal of ready, articulated theories. All the rest he sees as a matter for an empirical psychology of discovery, outside the normative realm of the logic of discovery. This represents an all-important shift in the problem of normative philosophy of science. *The term 'normative' no longer means rules for arriving at solutions, but merely directions for the appraisal of theories already there.*[29]

In other words, drawing a sharp distinction between the so-called contexts of discovery and of justification, Lakatos does not address himself to the problem of how new ideas or theories emerge or are produced. About the closest that he comes to recognizing the importance of the production process of new knowledges is his remark, in a footnote, that 'The actual hard core of a programme does not actually emerge fully armed like Athene from the head of Zeus. It develops slowly, by a long, preliminary process of trial and error. In this paper the process is not discussed'.[30] Lakatos shares the metaphor with Koyré, but is blocked by his philosophical prejudices from actually undertaking an analysis — like that of Koyré's — of the production process. It is not systematically discussed by him in 'this paper', or anywhere else for that matter, because it falls outside the domain of his project — in the domain of the so-called 'context of discovery', in fact. As such he dismisses the process whereby a hard core emerges on the grounds that it constitutes material for a causal, empirical study of 'psychological' phenomena, having no place in a rational, normative study of disembodied systems of thought.

I shall expose some of the incoherences of Lakatos' anti-psychologism below. For the present it need merely be noted that, having evaded the question of how new knowledges are produced, Lakatos has somehow to get rival research programmes onto the historical stage, as it were. He does so by methodological fiat. Thus

> One must never allow a research programme to become a *Weltanschauung* . . . . The history of science has been and should be a history of competing research programmes (or, if you wish, 'paradigms'), but it has not been and must not become a succession of periods of normal science: the sooner competition starts, the better for progress. 'Theoretical pluralism' is better than 'theoretical monism': on this point Popper and Feyerabend are right and Kuhn is wrong.[31]

Lakatos, in other words, instructs us to proliferate. We are to invent new, exciting and original ideas. As we have seen, for Popper this process 'seems to me neither to call for logical analysis nor to be susceptible of it'.[32] It is an act of 'poetic intuition'. For Feyerabend the sources of these new ideas can be dreams, fantasies, literature, fairy-tales — all those realms, in fact, which are dismissed and repressed in an overly narrow and dogmatic scientific education.[33] By delving into these 'non-rational' and esoteric domains we not only experience the exhiliration which comes from expanding the horizons of our awareness; we also contribute to the advance of scientific knowledge.

But do we? In fact the proliferation that Lakatos demands is supposed to generate a very particular kind of theory, viz. one that is a *competitor* or *rival* to the prevailing view. Obviously then not any fantastic idea can be put forward and defended on the grounds that it is contributing to the growth of scientific knowledge (although of course it may have other merits). Accordingly it is imperative that Lakatos, who claims to be concerned with ways of achieving scientific progress, explore the conditions under which, and the processes whereby, *rivals* are produced. If the history of science has, in fact, been a history of competing research programmes as Lakatos asserts, then the onus is on him to analyze the conditions under which they are produced, if only to ensure that the production of new knowledge may continue to occur. But instead of historical analysis he gives us methodological dicta; they are the masks behind which he hides his failure to engage with the historical process.

This point can be made in another way. At the start of this section I quoted Lakatos as commending Popper for developing a conception of rational choice that involved 'rules for the appraisal of ready, articulated theories'. Immediately thereafter I noted that Lakatos realizes that it takes time for scientists to build up even the hard core of a theory. While they are doing so, then, his (and Popper's) normative considerations are quite irrelevant to them. *They* do not have to choose between ready, articulated theories; they have to reject one theory and to commit themselves to a blossoming alternative. By restricting themselves to a rigid distinction between the context of discovery and the context of justification, Lakatos and Popper develop methodologies that have no bearing whatever on the decisions that need to be taken by those engaged in producing a new system of thought. Trapped by their philosophical prejudices, Lakatos and Popper take a spectator's, not a participant's perspective on the historical process. From that vantage point they have no hope of ever theorizing major historical changes as rational, and their prescriptions are irrelevant to those who are struggling to forge a new conceptual scheme.

3.5 *The End of Crucial Experiments — and its Aftermath*

As we have seen, although Lakatos does not explore the process whereby new theories are produced, his conception of 'revolutionary' change *presupposes* that quite particular kinds of relationship pertain between the old and the new. Not only does the latter emerge as a rival or competitor to the former: as it develops, Lakatos asserts that the two will (or ought to) intersect with each other. Their paths cross in 'crucial experiments' or, rather, they touch and can diverge again — thus preserving a measure of continuity. Lakatos writes

> Popper, as we have seen, did not explain some important aspects of *continuity* in the growth of science. But did we not go in our 'anti-falsificationist' approach so far to the other extreme that now we are bound to say that even the celebrated *'crucial experiments'* have no force to overthrow a research programme ?
> The answer is very easy. In the progress of science there is a *proliferation of competing research-programmes* . . . . As the rival research-programmes expand, they gradually encroach on each other's territory and the *n*th version of the first will be blantantly, dramatically inconsistent with the *m*th version of the second.[34]

However, Lakatos goes on to say, a defeat for the first programme in this skirmish is not reason enough for its immediate elimination or supersession. Any research programme is allowed a few such setbacks. All that is needed is for talented and imaginative scientists to modify the 'defeated' programme in a non-*ad hoc* way. If after repeated efforts they fail to do so, however, the experiment in question may, with hindsight, come to be regarded as crucial. It may take decades to appreciate this, though. There is no 'instant rationality' or quick-kill in science; rather, old and exhausted research programmes tend to wither away gradually.

The evaluation of move and countermove in this war of attrition is made 'horizontally', and the fortunes of each protagonist are intimately linked with those of the other. Thus

> The progress of one programme is a vital factor in the degeneration of its rival. If programme P_1 constantly produces 'novel facts', these, by definition, will be anomalies for the rival programme P_2. If P_2 accounts for these novel facts only in an *ad hoc* way, it is degenerating by definition. Thus the more P_1 progresses, the more difficult it is for P_2 to progress.[35]

On the other hand, even if P_2 is degenerating it is not necessarily irrational to continue working in it. For with luck and ingenuity scientists may be able to breathe life back into the most decrepit programme, setting it back on a 'progressive' path. Generally speaking it is difficult to decide when a research programme has degenerated hopelessly, or when one of the two or more rivals has finally achieved a decisive advantage over its competitors. The (naïve) falsificationist espoused a theory of rational progress which emphasized the importance of devastating crucial experiments as sounding the death knell of one set of rival programmes. Lakatos' methodology, as he puts it 'emphasizes long-extended theoretical and empirical rivalry of major research programmes, progressive and degenerating problemshifts, and the slowly emerging victory of one programme over the other'.[36] 'Rationality', he writes, 'works much slower than most people tend to think, and, even then, fallibly'.[37]

Lakatos, then, will not permit us to condemn as unreasonable scientists who stick to degenerating programmes. All that he requires is that the successes and failures of the rivals be publicly displayed so that the protagonists are able to judge whether or not a programme is progressing *at a particular point in time*. That assessment says nothing about future performance nor, correlatively, does it imply that one may not rationally work in a degenerating field. As Feyerabend puts it 'the methodology of research programmes provides standards that aid the scientist in *evaluating* the historical situation in which he makes his decisions; it does not contain rules that tell him what to do'.[38] That granted, the question which then arises is not whether rationality 'works' slowly or quickly, but whether it works at all. Lakatos appears indeed to have reacted so strongly against naïve falsificationism that it no longer seems to be possible rationally to reject a research programme, nor to condemn its adherents as unreasonable. Has his quest for continuity, then, led Lakatos into admitting that even crucial experiments 'have no force to overthrow a research programme'? Reply: in the methodology of research programmes the force, if any, which crucial experiments do have to eliminate a theory is not that derived from logic alone. Logic and experiment need to be supplemented by other pressures if we are ever to eliminate a system of thought once it has established itself.

What other factors 'encourage' or coerce scientists to change their allegiance from a degenerating research programme to its progressing rival? Lakatos suggests that it is psychological and sociological pressures that help to precipitate the transition. Elaborating in an entertaining way on some rather

LAKATOS' QUEST FOR CONTINUITY

brief remarks of his to this effect, Feyerabend spells out a possible scenario:

> Assume that the institutions which publicize the work and the results of the individual scientist, which provide him with an intellectual home where he can feel safe and wanted and which because of their eminence and their (intellectual, financial, political) pull can make him seem important adopt a *conservative attitude* towards the standards, they refuse to support degenerating research programmes, they withdraw money from them, they ridicule their defenders, they do not publish their results, they make them feel bad in any possible way. The outcome can be easily foreseen: scientists who are as much in need of emotional and financial security as anyone else, especially today, when science has ceased to be a philosophical adventure and has become a business, will revise their 'decisions', and they will tend to reject research programmes on a downward trend.[39]

Thus the ultimate victory of one research programme over another is the outcome of a combination of 'objectively' based judgements as to whether or not one of the rivals is progressing, along with a reluctance to allow any further leeway to a tottering, yet not necessarily defunct, system of thought.

It is important to be aware of how Lakatos arrived at this position. He found that naïve falsificationism, and other methodologies which he considered, severely restricted the scope of reason as an agent of progress in science. Accordingly, much of the actual history of science appeared to be shaped by irrational and non-rational forces. Lakatos was unwilling to accept this. If philosophers formulated theories of rationality in the light of which major past scientific achievements appeared to be the outcome of unreasonable behaviour, so much the worse for the philosophers. He insisted that it was their theories of rationality that needed to be modified, and not the heroes of science who needed to be condemned. Within this general framework it was Kuhn's response to naïve falsificationism that particularly disturbed him. As far as Lakatos was concerned, the 'instant rationality' of conjecture, refutation and theory elimination was as reprehensible as the instant non-rationality of a gestalt switch. But having emphasized continuity and tenacity, Lakatos found himself in the unenviable position of being unable to provide clearcut criteria for evaluating as unreasonable any behaviour whatsoever. It was as reasonable to stick to a new programme, surrounded by an ocean of anomalies, as it was to cling doggedly to a thoroughly decrepit system of thought on the grounds that it may just recover. In the last resort it was to psychological and social pressures, construed as non-rational, that Lakatos appealed to explain how it was that scientists gradually changed their allegiance from a degenerating research programme to a progressing rival.

Lakatos' difficulties bring home just how important the principle of empiricism, as Popper called it, is to a philosophical tradition that rigidly demarcates justification from discovery. In the empirical base of science Popper found that ultimate arbiter which could be used rationally to *eliminate* inadequate theories. Once that base is undermined, so too is the possibility of rationally rejecting a scientific theory and of replacing it with a better one. This does not simply mean that it becomes increasingly difficult to theorize revolutionary, discontinuous transitions as rational. It also involves a slide towards that form of relativism which asserts that 'anything goes', that there are not nor have there been rational grounds for eliminating any system of thought whatsoever. It is no wonder that Feyerabend describes Lakatos as an epistemological anarchist in disguise !

3.6 Digression: The Logical and The Socio-Psychological

The discussion above reveals that there is more than a little justification for Kuhn's claim that for both him and Lakatos the choice between competing systems of thought 'cannot be compelled by logic and experiment alone'.[40] Lakatos likes to stress that the reasons that he gives for the crumbling of a hard core are 'mainly logical and empirical',[41] whereas Kuhn's are predominantly socio-psychological. This one-sidedly overlooks the stress which Kuhn places on the build-up of empirical anomaly as one factor triggering crises within a paradigm. In Kuhn's theory a 'hard core' is put in jeopardy by an accumulation of negative instances, or by what we may call a logic of negative induction. For him, as for Lakatos, logic and experience, however, are not sufficiently persuasive to steer people across a revolutionary divide. He supplements them with gestalt switches and conversion experiences.

Lakatos locates the extraneous sources of pressure in the scientific establishment. By ostracizing those who stick to a degenerating research programme, its members effectively destroy such alternatives to their preferred system.

Actually Lakatos' remark that logic and experiment are mainly responsible for the demise of a research programme is best taken more as a pious act of faith than as a conclusion derived from historical analyses. For as I have stressed one of his aims is to formulate a theory of rationality which accounts for actual scientific change as being predominantly reasonable. Unfortunately, enterprises of this kind have brought little credit to those philosophers who have implemented them. Both Popper and Lakatos have rewritten (or 'rationally reconstructed') history to fit their methodological preconceptions — and have ended up caricaturing it. 'What Lakatos conceives as history is not history at all', writes Kuhn, 'but philosophy fabricating examples'.[42] To avoid this difficulty Lakatos has recently suggested that methodologies be treated as 'quasi-empirical' theories of history. Rather than rewriting history in accordance with their precepts, we should assume that, to a first approximation, the actual behaviour of the scientific elite over the last two centuries has been rational. Methodologies (i.e., theories of history) are then tested against that elite's assessment of specific achievements in science, and modified to accommodate 'anomalies' (i.e., judgements by the elite which would otherwise be regarded as lapses of rationality). By applying the methodology of research programmes to our methodologies treated as theories of history we can gradually evolve a theory of rational scientific behaviour, which reflects that of the scientific elite in question. The 'fit' will never be perfect, because non-rational factors will always play some role in shaping the basic value-judgements of scientists. However, the relative weight of such factors will be steadily diminished, at least in those contexts where the scientific community was not coerced by extra-scientific pressures to switch allegiance from one theory or research programme to another. What's more, the precise points at which 'socio-psychological' or 'external' accounts of scientific change need to be invoked will be thrown into relief.

Feyerabend has been particularly critical of this manoeuvre.[43] He has pointed out that it takes the excellence of the science of the past two centuries for granted, that it presupposes that the basic value-judgements of the scientific elite are uniform, and that it assumes that by and large their judgements are sound ones. However, notes Feyerabend, the common scientific wisdom is neither particularly common nor is it notably wise — as Lakatos implicitly concedes.[44]

For the present purpose I wish to stress a rather different, though related, aspect of Lakatos' project. I pointed out above that 'rational reconstructions' of science have caricatured actual scientific change, and have amounted to little more than what Kuhn called 'philosophy fabricating examples'. As I have suggested, the roots of this defect lie in Lakatos' and Popper's allegiance to a rigid dichotomy between the contexts of justification and of discovery. This leads them to construe their enterprise as an essentially normative one and, as such, logically independent of a socio-psychological study which explains people's actual beliefs and actions. Accordingly they write a fantasy-history which describes what real historical agents *would have done* had they behaved in accordance with a particular theory of rationality. Whether or not they did behave in accordance with, say, Popperian canons of rationality they (allegedly) take to be an entirely separate matter. As Lakatos puts it, the

> problem of appraisal is completely different from the problem of why and how new theories emerge. Appraisal of change is a normative problem and thus a matter for philosophy; explanation of change (of actual acceptance and rejection of theories) is a psychological problem. I take this Kantian demarcation between the 'logic of appraisal' and the 'psychology of discovery' for granted.[45]

Now if this dichotomy had any substance to it, it really would not matter if there turned out to be a colossal mismatch between the philosopher's fantasy-history and the historians real history. After all, they are supposed to be logically independent of each other. Yet they also appear to be intimately linked with each other. For the philosopher's theories of rationality lay down canons of reasonable *behaviour*, and the actual behaviour of scientists can be assessed — from a rational point of view — in their light. It was precisely this relationship that stimulated Lakatos to look more closely at history. Finding that the behaviour of scientists as described by historians was irrational in terms of naïve falsificationism, he set out to develop a more sophisticated conception of reason, which could account for more of the actual practice of science as rational.

The philosophy of science, writes Lakatos, 'gives us guidance as to when the acceptance of a scientific theory is rational or irrational'. What he does not appreciate is that to claim that someone's acceptance of a scientific theory is rational *is* to explain that person's behaviour. People frequently accept a theory because it is reasonable to do so: the development of Lakatos' methodology discussed in this section assumes precisely that the scientific elite of the last two centuries have made reasonable moves in the game of science most of the time. Under such circumstances, in answering a question of fact — why someone actually accepted a theory — one also answers a question of value, namely whether that view was the reasonable one to adopt. One *both* explains and evaluates (from a rational point of view) the state of affairs constituted by that person's belief in the theory in question. This is not to say, of course, that people always accept theories on rational grounds. It is merely to concede that they often do so. People may not be wholly rational all the time; but Lakatos for one, like most of us, assumes that they can be at least partly rational a good deal of the time. When the acceptance of a scientific theory is guided (perhaps only in part) by irrational considerations, an explanation of that event as reasonable must be supplemented, or replaced, by one which appeals to other pressures. Although he does not realize it, this is the rationale for Lakatos' repeated insistence that an 'internal' account of historical change is primary, but that it can leave a residue which has to be explained in terms of 'external' (i.e., non-rational) factors.

Lakatos likes to point out that his approach is normative, whereas Kuhn's is socio-psychological. Thus baldly stated, this is a false dichotomy. Like Popper, Lakatos thinks that evaluative logical terms like 'consistent' and 'contradictory' pertain soley to the relationships between statements or propositions (i.e. to items that can be true or false). However, to describe two propositions as inconsistent with each other is to imply that it would be unreasonable or inconsistent *to believe or to accept* both of them. In other words what reason evaluates is not just a statement or theory but people's beliefs and actions.[46] This simple truism, which is obscured beneath layers of philosophical prejudice, has been emphasized and argued for by Edgley.[47] He has suggested that reason is a faculty whose essential function is the normative one of guiding other faculties which are subordinate to it. In particular it is a faculty which brings pressure[48] to bear on people's psychological states, like belief and theory acceptance, and on their actions. Accordingly any rigid dichotomy between logic and psychology is incoherent — as the work of Lakatos himself clearly shows.

The contradictions generated by a failure to recognize this point are also apparent in the paper by Lakatos and Zahar on the Copernican Revolution. It is allegedly addressed to the question of why Copernicus' research programme superseded Ptolemy's. This sounds as if the authors' aim is to provide an historical explanation of scientific change. However, they insist that this is not so. They are only concerned, they say, with the normative aspect of the problem; why people actually became Copernicans does not concern them. But this does not stop them criticizing the accounts of the Copernican Revolution given by Kuhn and others who *are* concerned to explain how that event actually took place ! In fact Lakatos and Zahar insist that 'the Copernican Revolution can be explained as rational on the basis of the methodology of scientific research programmes' and that it 'explains Copernicus' achievement as constituting genuine progress compared with Ptolemy'.[49] So they *are* concerned with why Copernicus espoused a heliostatic view, and they are arguing that he was rational to do so. It is reassuring to have Lakatos admit that he takes a dichotomy between evaluative matters of logic and empirical matters of psychology for granted. At least we are then able to understand why he unwittingly violates it at every turn.

3.7 Revolutions without Rivals

When I introduced the discussion of Lakatos' theory of scientific revolutions I remarked that he did not systematically differentiate between (at least) two conceptions of novelty. On one conception, adjustments to a research programme in the light of anomaly are not *ad hoc* if they lead to the prediction of novel facts which are forbidden by a rival programme. Here it is the fate of an entire research programme which is in the balance, and its progress in comparison with a rival must be gauged. But there is another conception of novelty introduced by Lakatos in the context of his discussion of sophisticated falsificationism. Here the assessment of novelty is made independently of the presence of a rival programme. And actually even the earlier state of development of the programme is only marginally relevant to it. For example, confirming evidence for the latest member in the series of theories need not disconfirm its predecessor; all that Lakatos asks is that the new prediction be 'unexpected'. On this view *temporal* novelty is virtually sufficient to establish that a particular move within a research programme is (theoretically) progressive.

As I have stressed, Lakatos tends indiscriminately to merge these conceptions of novelty. For example, he writes

I originally defined a prediction as 'novel', 'stunning', or 'dramatic' if it was inconsistent with previous expectations, unchallenged background knowledge and, in particular, if the predicted fact was forbidden by the rival programme. The best novel facts were such which might never have been observed if not for the theory which anticipated it (*sic*).[50]

This is a ghastly muddle. Facts inconsistent with previous expectations need not emerge against the backdrop of a rival; facts forbidden by a rival programme can only be detected if a competitor is available. And a fact that might never have been observed but for the theory that anticipated it is not necessarily inconsistent with previous expectations or a rival programme. Their level of articulation or domain of application may be such as to make it impossible to predict the fact in question.

The conception of novelty expressed in the second sentence of this quotation is essentially that which Zahar attributes to Lakatos. Zahar notes that Lakatos likes to treat the return of Halley's comet as a typical novel fact predicted by Newton's research programme. Rather simplistically, he attributes to Lakatos the view that 'the discovery of any new type of fact is the discovery of a novel fact'.[51] But, he goes on to say, the identification of novelty with temporal novelty is unduly narrow: it excludes already known facts as legitimate confirming evidence for a new theory. To accommodate them Zahar proposes that a fact be considered novel if it does not 'belong to the problem-situation which governed the construction of the hypothesis'.[52] On this view any new experimental results obtained with a theory will be novel, as well as facts already discovered but not consciously incorporated into its original design.[53]

Lakatos regards Zahar's conception of novelty to be 'a new criterion of scientific progress — a criterion which is a very important amendment to that provided by the methodology of scientific research programmes'.[54] It is, however, more than an amendment: it is in flagrant contradiction with some of the most basic tenets of that methodology. More particularly, Lakatos has characterized a research programme as stagnating 'if its theoretical growth lags behind its empirical growth, that is, as long as it gives only *post-hoc* explanations either of chance discoveries or of facts anticipated by, and discovered in, a rival programme'.[55]

However, the facts which Zahar admits as novel are precisely facts already available when a theory is formulated, and may well be facts discovered in a rival programme. Thus a programme judged to be degenerating on Lakatos' original conception of novelty is judged to be progressing on Zahar's new, and so-called 'amended', conception !

Needless to say these lacunae, ambiguities and contradictions in the theory of novelty and scientific progress are never confronted or resolved by Lakatos. Nevertheless there is a general trend in the 'development' of his conception of novelty, namely, a steady decline in the emphasis which is placed on the hard core of research programmes in assessing progress or degeneration. When we remember that novelty was originally introduced within the context of sophisticated falsificationism, which assessed progress 'vertically' within a series of theories with a shared hard core, we can appreciate that this risk was always there. As reference to the hard core becomes less significant in judging progress, so too does the demand for the proliferation of programmes with rival hard cores become less urgent. Revolutionary transformations are no longer conceived of primarily as protracted wars of attrition between two or more competing programmes, and the idea that one programme gradually overtakes-in-progress its rivals recedes into the background. If only by default it is 'assumed' that there is a continuity of fundamentals across a revolutionary divide. More precisely, perhaps, the possibility of there not being a continuity at the level of basic postulates during revolutionary upheavals is never confronted. One can hardly conclude that progress required, or may require, the elimination or supersession of one research programme by another if the hard core of the programmes is more or less irrelevant to one's assessment of progress ! In effect, then, Lakatos and Zahar leave us with a theory of revolution without rivals, in the formulation of which the conflict over hard cores is, at best, of only indirect and marginal importance.

The concrete manifestations of these tendencies are all too apparent in the historical analyses of revolutionary episodes made by these authors. In two lengthy and intricate papers Zahar has addressed himself to the question of why Einstein's relativistic programme superseded Lorentz's classical programme — yet he never spells out the hard core of Einstein's programme. This, I am suggesting, is not merely an accidental oversight but a direct consequence of Zahar's conception of novelty and his criteria for scientific progress. Similarly in Lakatos and Zahar's analysis of the Copernican Revolution, an absolutely fundamental issue that was at stake — the question of the Earth's motion — is treated as if it were of only secondary importance. Feyerabend suggests that this leads the authors to assess as progressive a programme which was, in fact, degenerating

The research programme Copernicus wants to overthrow (which I shall call A) contains a cosmology, a dynamics as well as certain methods for calculating the paths of the planets P_A. His own research programme, K, contains a new and unified method for calculating the paths of the planets, P_K. Lakatos and Zahar compare P_A and P_K only. They write: 'Ptolemy's programme always dealt with facts *post hoc* and hence always degenerated. Copernicus' model of the *Commentariolus* constituted dramatic progress'. The 'dramatic progress' turns into stagnation if we compare A and K which are the research-programmes at issue (all dynamical problems have to be taken care of *post-hoc*; the same is true of optical problems). Concentrating on P_A and P_K also means adopting the instrumentalism of the Ptolemaians which Copernicus tried to overcome.[56]

The fine structure of the argument criticized by Feyerabend

reflects the insiduous effects on Lakatos' and Zahar's (hereafter cited as LZ) work of formulating conceptions of novelty and of progress within a programme which entail little or no reference to its rival. The aim of their paper is to show that it was rational to adopt the Copernican programme in 1543, when Copernicus' *De Revolutionibus* was published. From the outset the authors push the hard cores of the two programmes into the background

> Both *programs* branched off from the Pythagorean–Platonic program whose basic principle was that since heavenly bodies are perfect, all astronomical phenomena should be saved by a combination of as few uniform circular motions (or uniform spherical rotations about an axis) as possible. This principle remained the cornerstone of the heuristic of both programs. This proto-program contained no directives as to where the center of the universe lies. The heuristic in this case was primary, the 'hard core' secondary.[57]

Having suppressed by methodological fiat the issue of geocentrism versus heliocentrism, LZ plunge into an evaluation of the development of the geostatic programme. They claim that the introduction of eccentrics and epicycles ran counter to the Platonic heuristic and that the use of the equant by Ptolemy amounted to abandoning it. What's more LZ point out that 'despite all these violations of the Platonic heuristic, the geostatic program remained empirically *ad hoc*, that is, it always lagged behind the facts'.[58] Copernicus got rid of the equant, producing a system which was heuristically superior to Ptolemy's and which in 1543 'had *immediate* support from facts which were novel in Zahar's sense'.[59] Conclusion: 'Zahar's account thus explains Copernicus' achievement as constituting genuine progress compared with Ptolemy'.[60]

Without any shadow of doubt this is one of the most incoherent and intellectually irresponsible accounts of the Copernican Revolution that one is ever likely to come across. I shall consider just two of its more glaring defects.

Firstly, the introduction of eccentrics, epicycles and the equant is said by LZ to run counter to the Platonic heuristic and they imply that this is because these constructions involved motion about an empty point in space.[61] In addition, motion about the equant was non-uniform. However, eccentrics and epicycles are quite consistent with the heuristic as specified by LZ, which states merely that astronomical phenomena should be saved by combinations of uniform circular motions. This is compatible with some motions being about an empty point. As a matter of fact LZ tacitly assume that epicyclic motion is *not* heuristically degenerate when they appraise Copernicus' theory. They recognize that Copernicus resorted to epicycles, but say nothing at all about their being heuristically degenerate when used in the heliostatic system ! They stress only Copernicus' heuristic superiority over Ptolemy, referring to the abolition of the equant.

The second gross defect in LZ's account of the Copernican Revolution to which I wish to draw attention also involves the unwitting application of a double standard. The conception of novelty, which LZ primarily use to dismiss the Ptolemaic system, is one which takes a research programme to be empirically degenerate if it absorbs facts already known. But the conception of novelty which they use to defend as rational a switch to Copernicanism in 1543, is one which takes a research programme to be progressive if it *can* account for facts already known, thought not built into the original conception of the theory. LZ make no attempt whatsoever to evaluate the progressiveness or otherwise of the Ptolemaic system in the light of this quite different criterion of novelty.

Not content with this, LZ make the fantastic claim that the fact that planets have stations and retrogressions was a Zahar-type novel fact in favour of Copernicanism. This is ridiculous. It is absurd to suggest that Copernicus was not aware of the irregularities in planetary motion when he formulated his astronomical theory, and that stations and retrogressions were an 'unexpected present', or an 'unintended by-product' of his hypothesis.

How can LZ possibly have overlooked this ? To illustrate 'the corroborating strength' of this and other facts they propose that we perform an historical thought experiment. We are asked to imagine that in about 1520 all that was known about the heavens was that the Sun and the planets moved periodically with respect to the Earth. Records, however, were so scanty that planetary stations and retrogressions had never been experimentally detected. Two astronomers, both of whom adopt the Platonic heuristic, are postulated. Astronomer X adopts the heliocentric model, astronomer Y adopts the corresponding geocentric mode. Then X's theory would have been dramatically confirmed by subsequent observations of retrograde motion, whereas Y's could only have been saved in an *ad hoc* way. 'Zahar's account,' concludes Lakatos, 'thus explains Copernicus' achievement as constituting genuine progress compared with Ptolemy'.[62]

But of course it does not do this at all. At best it explains why it was rational for some hypothetical ignoramus X to adopt a Copernican system. And anyway in the thought experiment it is asserted by LZ that the 'novel facts' in favour of X's view were observed *after* he had put forward his theory. But when they listed four novel facts in favour of the heliostatic view in 1543 they specifically said that these facts 'were previously known'. The thought experiment uses Lakatos' original conception of novelty to 'illuminate' an argument which relies on Zahar's quite different conception.

Lakatos and Zahar's attempt to show that it was rational to adopt Copernicanism in 1543 has plainly had disastrous consequences. In making their case they have ignored the most rudimentary demands of good scholarship. Any account of the Copernican Revolution that treats the issue of the movement of the Earth as secondary is surely doomed to failure.[63] Moreover, in their case such an approach is the direct outcome of a tendency to play down the revolutionary aspects of scientific change. By pushing the 'hard cores' of rival programmes into the background a spurious continuity is aprioristically imposed on the historical process. This may satisfy Lakatos' deepfelt need for gradual change, but it does so at the expense of any adequate appreciation of the depth and breadth of revolutionary upheavals.

3.8 Concluding Remarks

Summarizing what he takes to be some of his major achievements, Lakatos has said that he hopes he has shown that

> theories of instant rationality — and instant learning — fail . . . rationality works much slower than most people tend to think, and, even then, fallibly. Minerva's owl flies at dusk. I also hope I have shown that the *continuity* in science, the *tenacity* of some theories, the rationality of a certain amount of dogmatism, can only be explained if we construe science as a battleground of research programmes rather than of isolated theories.[64]

This project, it will be remembered, is developed within the framework of a rigid dichotomy, sanctified by Popper, between normative matters of logic, on the one hand, and empirical matters of psychology, on the other. Accordingly, with Popper, Lakatos does not seriously address himself to the question of how new knowledges are produced or, rather, he simply asserts that their production is non-rational, and falls within the 'context of discovery'. His basic aim is then to develop a theory of rational progress which is limited to the 'context of justification', when a choice has to be made between fully articulated theories or research programmes.

What Lakatos objects to is Popper's attempt to force scientists to make relatively quick, irreversible decisions to eliminate scientific theories. In the interests of continuity and tenacity the transition from one research programme to another is considered to be a gradual process. In fact, Lakatos argues that more often than not it is reasonable to retain one's allegiance to a research programme virtually indefinitely. He thus drastically softens the conditions for the rational rejection of a system of thought, playing down the importance of anomaly and allowing more scope to the imagination and ingenuity of scientists than Popper ever did. As a result Lakatos is only able to 'kill off' a programme by invoking socio-psychological pressures. Alternatively, he unwittingly imposes a spurious continutiy on revolutionary transformations by constructing a conception of progress which refers only obliquely to the hard core of a programme.

For Lakatos reason demands peaceful co-existence with little more than the odd border-skirmish on the 'battleground of research programmes'. Whereas Popper tends to emphasize struggles between *rivals* of which only the fittest survive, in Lakatos the 'rivals' begin to look more and more like equally reasonable alternatives. Détente, in other words, has its disadvantages. Specifically, it protects as reasonable an outworn system of thought on the grounds that it may just manage to recover its former magnificence. Lakatos' tolerance conceals a conservatism, then, which springs directly from his unwillingness to reject research programmes, and to replace them with better ones. This tendency in his work reaches fruition in Feyerabend's epistemology, to which I shall now turn. Here it is argued that 'pure tolerance' is the only rational attitude to adopt towards systems of thought, both old and new. Lakatos, in other words, stands midway between Popper and Feyerabend, and it is as such that I think his contribution to current debates in the philosophy of science is best understood and assessed.

Notes

1. P. Feyerabend, 'Imre Lakatos', *British Journal for the Philosophy of Science*, 26 (1975), 1.
2. I. Lakatos, 'Falsification and the methodology of scientific research programmes', in I. Lakatos and A. Musgrave (eds), *Criticism and the Growth of Knowledge* (Cambridge University Press, Cambridge, 1970), p.174.
3. K.R. Popper, *Conjectures and Refutations* (Routledge & Kegan Paul, London, fifth edition, 1974), pp. 51, 313, original emphasis suppressed.
4. I have criticized this tactic in J. Krige, 'Popper's epistemology and the autonomy of science', *Social Studies of Science*, 8 (1978), 287 – 307.
5. K.R. Popper, *op.cit.* p.54.
6. T.S. Kuhn, *The Structure of Scientific Revolutions* (University of Chicago Press, Chicago, second edition, 1970), p.77.
7. Lakatos further differentiates 'naïve' falsificationism from 'dogmatic' falsificationism. The essential difference between the two is that, whereas the latter regards science as resting on an irrefutable bedrock of sense experience, naïve falsificationism sees it as being built on a set of fallible, theory-laden observation statements. Naïve falsificationism takes science to be an edifice erected on piles driven into a swamp. Its basic statements are adopted by the free decision of the members of the scientific community. They can be revised when that community is no longer satisfied that they are capable of bearing the weight of the structure erected on them. The principle of empiricism requires that scientists decide which statements are 'theoretical' and which 'observational' in a particular test situation, and then decide on the truth-value of the latter. According to Lakatos, Popper has never held the 'dogmatic' view about the observational basis in science, although some of his interpreters have attributed it to him. The decisions about the (fallible) observational base must also be taken by the 'sophisticated falsificationist' (see below). As sophisticated falsificationists 'We cannot avoid the decision which sort of propositions should be the "observational" ones and which the "theoretical" ones. We cannot avoid either the decision about the truth-value of some "observational propositions"' (see I. Lakatos, in I. Lakatos and A. Musgrave, *op.cit.* p.127).
8. I. Lakatos, in I. Lakatos and A. Musgrave (eds), *op.cit.* p.177. Kuhn, of course limited continuity to periods of 'normal' science. Lakatos, as we shall see, tries to extend it to cover scientific revolutions as well.

9   K.R. Popper, *The Logic of Scientific Discovery* (Hutchinson, London, 1959), p.50.
10 *Ibid.* p.80.
11 *Ibid.* p.84.
12 *Ibid.* pp.82–3.
13 *Ibid.* p.83.
14 *Ibid.* p.50.
15 I. Lakatos, in I. Lakatos and A. Musgrave (eds), *op.cit.* p.117.
16 *Ibid.* p.118.
17 *Ibid.* p.132.
18 *Ibid.* p.135
19 P.K. Feyerabend, 'On the critique of scientific reason', in R.S. Cohen, P.K. Feyerabend and M.W. Wartofsky (eds), *Boston Studies in the Philosophy of Science*, vol.39 (Reidel, Dordrecht, Holland, 1976), p.120.
20 P.K. Feyerabend, 'Imre Lakatos', *op.cit.* p.2.
21 I. Lakatos, in I. Lakatos and A. Musgrave (eds), *op.cit.* p.134. Popper similarly required that for a new theory to be seriously entertained it merely needed to predict a new and unexpected fact, and that it only had to be empirically successful 'reasonably often' (see Chapter 2).
22 I. Lakatos, 'Popper on demarcation and induction', in P.A. Schilpp (ed) *The Philosophy of Karl Popper* (Open Court, Illinois, 1974), pp.242–3.
23 I. Lakatos, 'Criticism and the methodology of scientific research programmes', *Proceedings Aristotelian Society*, 69 (1968), 167.
24 P.K. Feyerabend, 'Zahar on Einstein', *British Journal for the Philosophy of Science*, 25 (1974), 25.
25 P.K. Feyerabend, in R.S. Cohen, P.K. Feyerabend and M.W. Wartofsky (eds), *op.cit.* pp.120–1.
26 *Ibid.* p.128.
27 I. Lakatos and E. Zahar, 'Why did Copernicus' research program supersede Ptolemy's ?', in R.S. Westman (ed), *The Copernican Achievement* (University of California Press, Berkeley, 1975), p.369.
28 *Ibid.* p.375.
29 I. Lakatos, in P.A. Schilpp (ed), *op.cit.* p.242.
30 I. Lakatos, in I. Lakatos and A. Musgrave (eds), *op.cit.* p.133.
31 *Ibid.* p.155, original emphasis suppressed.
32 K.R. Popper, *The Logic, op.cit.* p.31.
33 P.K. Feyerabend, 'On the improvement of the sciences and arts, and the possible identity of the two', in R.S. Cohen and M.W. Wartofsky (eds), *Boston Studies in the Philosophy of Science*, vol. 3 (Reidel, Dordrecht, Holland, 1967), p.405.
34 I. Lakatos, 'Criticism and the methodology of scientific research programmes', *op.cit.* p.176.
35 I. Lakatos, 'History of Science and its rational reconstructions', in R.C. Buck and R.S. Cohen (eds), *Boston Studies in the Philosophy of Science*, vol. 8 (Reidel, Dordrecht, Holland, 1971), p.125.
36 *Ibid.* p.105.
37 I. Lakatos, in I. Lakatos and A. Musgrave (eds), *op.cit.* p.174.
38 P.K. Feyerabend, in R.S. Cohen, P.K. Feyerabend and M.W.

Wartofsky (eds), *op.cit.* p.121.
39 *Ibid.* p.124.
40 T.S. Kuhn, 'Notes on Lakatos', in R.C. Buck and R.S. Cohen (eds), *Boston Studies in the Philosophy of Science*, vol. 8, (Reidel, Dordrecht, Holland, 1971), p.144.
41 I. Lakatos, in I. Lakatos and A. Musgrave (eds), *op.cit.* p.134.
42 T.S. Kuhn, 'Notes', *op.cit.* p.143.
43 P.K. Feyerabend, in R.S. Cohen, P.K. Feyerabend and M.W. Wartofsky (eds), *op.cit.* Section 2.
44 Lakatos realizes that it can err in particular cases, as Feyerabend points out. When that happens, the philosopher's universal, statute law should be invoked to overrule the scientific elite's case law as follows: 'The statute law approach should become much more important when a tradition degenerates or a new bad tradition is founded. In such cases statute law may thwart the authority of the corrupted case law, and slow down or even reverse the process of degeneration.' (I. Lakatos in R.C. Buck and R.S. Cohen (eds), *op.cit.* p.122). And so, once again, Lakatos appeals to a form of 'external' coercion to ensure that degenerating research programmes do, in fact, crumble, rather than staggering on indefinitely.
45 I. Lakatos and E. Zahar, *op.cit.* p.355.
46 As Lakatos implicitly concedes, *Ibid.* p.356.
47 R. Edgley, *Reason in Theory and Practice* (Hutchinson, London, 1969), Chapter 3.
48 I have discussed the significance of this metaphor in J. Krige, 'A critique of Popper's conception of the relationship between logic, psychology and a critical epistomology', *Inquiry* 21 (1978), 313–35.
49 I. Lakatos and E. Zahar, *op.cit.* pp.368, 380.
50 *Ibid.* p.375.
51 E.G. Zahar, 'Why did Einstein's programme supersede Lorentz's ? (I)', *British Journal for the Philosophy of Science*, 24 (1973), 101.
52 *Ibid.* p.103, original emphasis suppressed.
53 Zahar's definition of novelty is actually more complicated than this. He posits a succession of theories T_1 and T_2 in the same research programme. e_1 and e_2 are anomalies faced by T_1. T_2 is specifically evolved in order to account for e_1. If it is then found that T_2 can also account for e_2, then e_2, though not e_1, is novel with respect to T_2. These qualifications are ignored in the paper by Lakatos and Zahar on the Copernican Revolution to be discussed below. Musgrave also overlooks them in an article which may well have been read in proof by Zahar. See A. Musgrave, 'Logical versus historical theories of confirmation', *British Journal for the Philosophy of Science*, 25 (1974)1–23.
54 I. Lakatos and E. Zahar, *op.cit.* p.375.
55 I. Lakatos, in R.C. Buck and R.S. Cohen (eds), *op.cit.* p.100.
56 P.K. Feyerabend, 'Zahar on Einstein', *op.cit.* p.27.
57 I. Lakatos and E. Zahar, *op.cit.* pp.370–1.
58 *Ibid.* p.373.
59 *Ibid.* p.376.
60 *Ibid.* p.380.

61 This is explicitly stated by Zahar in his 'Why did Einstein's programme supersede Lorentz's ? (II)', *British Journal for the Philosophy of Science*, 24 (1973), 239.
62 I. Lakatos and E. Zahar, *op.cit.* p.380.
63 I shall substantiate this claim in Chapter 5.
64 I. Lakatos, in I. Lakatos and A. Musgrave (eds), *op.cit.* pp.174–5.

4 FROM DISCONTINUITY TO DADAISM

SUSSEX University: the start of the Autumn Term, 1974. There was not a seat to be had in the biggest Arts lecture theatre on campus. Taut with anticipation, we waited expectantly and impatiently for the advertized event to begin. He was not on time — as usual. In fact rumour had it that he would not be appearing at all, that illness (or was it just ennui ? or perhaps a mistress ?) had confined him to bed. But just as we began sadly to reconcile ourselves to the idea that there would be no performance that day at all, Paul Feyerabend burst through the door at the front of the packed hall. Rather pale, and supporting himself on a short metal crutch, he walked with a limp across to the blackboard. Removing his sweater he picked up the chalk and wrote down three questions one beneath the other: What's so great about knowledge ? What's so great about science ? What's so great about truth ? We were not going to be disappointed after all !

During the following weeks of that term, and for the rest of his year as visiting lecturer, Feyerabend demolished virtually every traditional academic boundary. He held no idea and no person sacred. With unprecedented energy and enthusiasm he discussed anything from Aristotle to the Azande. How does science differ from witchcraft ? Does it provide the only rational way of cognitively organizing our experience ? What should we do if the pursuit of truth cripples our intellects and stunts our individuality ? Suddenly epistemology became an exhilarating area of investigation.

Feyerabend created spaces in which people could breathe again. He demanded of philosophers that they be receptive to ideas from the most disparate and apparently farflung domains, and insisted that only in this way could they understand the processes whereby knowledge grows. His listeners were enthralled, and he held his huge audiences until, too ill and too exhausted to continue, he simply began repeating himself. But not before he had brought the house down by writing 'Aristotle' in three-foot high letters on the blackboard and then writing 'Popper' in tiny, virtually illegible letters beneath it !

In fact Popper and Lakatos have been two of the most important influences on Feyerabend's philosophical development. Popper, for all his emphasis on being open-minded and critical, has not been particularly receptive to Feyerabend's work, however. 'As far as my former pupil Feyerabend is concerned', he wrote recently, with breathtaking arrogance and complacency, 'I cannot recall any writing of mine in which I took notice of any writing of his'.[1] Lakatos has treated Feyerabend more kindly: he is, says Lakatos, 'our brilliant leading cultural relativist'.[2] As we saw in Chapter 3, Feyerabend has repaid him a hundredfold, with exaggerated and to my mind somewhat misplaced adulation.

Without undue distortion we may regard Popper and Lakatos as philosophers of continuous change, and Feyerabend as a philosopher of discontinuity. His awareness of the possibility of this kind of transition is founded on his recognition of certain historical facts which cannot be dissolved by any amount of philosophical bluster. In particular, he stresses that at the dawn of the seventeenth century there were several different cognitive schemes prominently available for organizing one's experience of the world. To mention three: (i) the comprehensive, complex and immensely powerful edifice of Aristotelian thought; (ii) a cluster of beliefs and practices pertaining to witchcraft, magic and astrology which were linked in complex ways to Hermeticism and Neo-Platonism; and (iii) the new mechanical philosophy of Galileo and Descartes. Of these, the third triumphed and science as we know it today is its heir. Aristotelianism is no longer regarded to be a viable system of thought and, in Western cultures at least, magical practices are peripheral and tolerated only so long as they remain within carefully specified limits.

Feyerabend is emphatic that those who are committed to progress and to understanding the growth of scientific knowledge should confront and engage with transitions of this kind. To do so is to recognize firstly that, insofar as scientific theories make assertions about what exists, the replacement of one conceptual scheme by another can be accompanied by a complete replacement of the ontology (and even of the formalism) of the one theory by that of the other.

These changes can be so far-reaching that they permeate down to the ordinary descriptive language of everyday speech. Typically

> Descartes' materialism played a very decisive role in discrediting the belief in spatially localizable spirits. The language of demonic influences was no part of the new conceptual scheme that was created in this manner. It was for this reason that a reformulation was needed, and a reinterpretation of even the most common 'observational' statements.[3]

However, discontinuous changes need not merely involve a replacement of one ontology by another; they can also involve a change in the standards of adequate explanation. For example, whereas the Aristotelians regarded mathematics as being of merely secondary importance for grasping the fluid structure of the material world (because it seemed to deal only with timeless, unchanging 'Platonic' entities), Galileo insisted that

> Philosophy is written in this grand book, the universe, which stands continually open to our gaze. But the book cannot be understood unless one first learns to comprehend the language and read the letters in which it is composed. It is written in the language of mathematics, and its characters are triangles, circles, and other geometric figures without which it is humanly impossible to understand a single word of it; without these, one wanders about in a dark labyrinth.[4]

As a matter of fact Galileo not only rejected Aristotelian standards, he also scornfully dismissed those canons of explanation bequeathed to him by the Hermetic tradition. Discussing Kepler's views that the tides were caused by lunar attraction, he remarked with astonishment that 'Despite his open and acute mind, and though he has at his fingertips the motions attributed to the earth, he has nevertheless lent his ear and his assent to the moon's dominion over the waters, to occult properties, and to such puerilities'.[5] Rather than appealing to sympathies and antipathies, Galileo provided a mathematical and mechanical account of the tides. The theory was in empirical difficulty from the start.[6] However, as Drake points out, 'To Galileo as a physicist, the purely mechanical basis of his explanation was decisive in its favour'.[7]

For Feyerabend, then, discontinuity both in ontology (and observation language) and in standards of (rational) explanation is an historical fact, which philosophers of science ignore at their peril. The comment by Drake that I have just quoted suggests one reason why this should be so. It suggests that a *prior* metaphysical commitment to a mechanistic (and mathematical) account of natural phenomena provided Galileo and his supporters with the psychological crutch which they needed to cling to a 'refuted' theory. According to Feyerabend such commitments have an essential role to play during periods of discontinuous transition. 'Progress' across a revolutionary divide is not achieved by slavishly implementing a set of methodological rules which are supposedly constitutive of rational behaviour. To get anywhere at all, the agents of change — precisely those agents whom we *now* admire as being paradigms of 'rationality' and 'objectivity' — had to follow their 'subjective', 'natural' inclinations. They had to be immensely flexible and ruthlessly opportunistic.

Developing these ideas Feyerabend has suggested that the lesson to be learnt from a careful study of historical events like the Scientific Revolution is that 'progress' can only be achieved in some situations by being an epistemological anarchist or, preferably, a Dadaist. Dadaists, he tells us, believe that life will only be worthwhile when we stop taking it too seriously, and free ourselves from a puritanical and dedicated search for 'truth' and 'justice'. They have no single goal, and the goals they do have vary with their inclinations and with the contingencies of the moment. To impress a mistress a Dadaist may vociferously defend the *status quo*; to rouse an angry and frustrated mob the same Dadaist may unmercifully criticize the establishment. Reasoning is simply a game which the Dadaist plays, preferably with those who are naïve enough to believe that we can get closer to the truth by attending to the content of arguments and by discussion.

Now, if this is Dadaism, Galileo was no Dadaist. At about twenty-five years of age he came to the conclusion that Aristotelian physics was completely untenable, and he extolled the virtues of reason and observation in the search for truth.[8] He spent the rest of his life working out the implications of these views. In the teeth of hostile opposition from the universities and from powerful factions within the church, he fought for what he believed was right. His determination brought him to his knees before the Inquisition, an old man compelled to deny his allegiance to Copernicanism. Even then it is said that as he stood up Galileo defiantly stamped one foot and muttered 'And yet it moves'. This anecdote may be apocryphal but it remains as a tribute to his legendary stubborness and courage. Without his determination to pursue and to defend the truth come what may, it is arguable whether Galileo would have been sufficiently motivated to attack the entire edifice of Aristotelian thought and those institutions whose authority rested on it.

There is thus something strikingly paradoxical about Feyerabend's views on how the growth of knowledge has been, and should be, achieved. Put bluntly, it is this: how is it possible for him to take Galileo's behaviour as typical of what is required to achieve such growth, sensitively exploring many of its nuances, and yet to end up defending — perhaps only half-seriously — epistemological anarchism or Dadaism ? It is to the resolution of this paradox that the argument in this chapter is primarily directed.

It is my contention that it is Feyerabend's way of theorizing revolutionary transitions that propels him from an awareness of discontinuity into the defence (with tongue in cheek ?) of Dadaism. One of the most important features of his work is his recognition of what he calls the 'uneven development' of science. As I noted in the previous chapter, he stresses (Ref. 19) that when a new theory enters the scene it is conceptually and empirically inferior to an established rival. Methodologies that adopt a rigid demarcation between the contexts of discovery and of justification and that assess 'ready, articulated theories' (Lakatos) are, as I pointed out in Section 3.4, of no value in such situations. As Feyerabend puts it, they 'project all the various elements of science and the different historical strata they occupy on to one and the same plane, and proceed at once to render comparative judgements. This is like arranging a fight between an infant and a grown man, and announcing triumphantly, what is obvious anyway, that the man is going to win'[9] However, as far as I am concerned Feyerabend's alternative is equally flawed. For even while rejecting the dichotomy between the contexts of discovery and of justification, Feyerabend borrows from it the notion that the production of new ideas is an essentially arbitrary and non-rational process. In what follows I shall argue the he accordingly provides an intrinsically defective and one-sided account of discontinuous transitions — and lands up in the dead-end of Dadaism.

4.1 Against Monomania

I pointed out in Chapter 1 that Kuhn's historical studies had led him to conclude that the emergence of new scientific theories involved 'destructive changes in beliefs about nature'. He went on to say that 'though logical inclusiveness remains a permissible view of the relation between successive scientific theories, it is a historical implausibility' (cf. Chapter 1, Ref. 20). These considerations stimulated his, and Feyerabend's, interest in discontinuous transitions. Valuable insights into the way in which Feyerabend approaches such changes can be gained by looking briefly at some of his earlier work.

One of Feyerabend's central concerns is to demolish the view that genuine knowledge is embodied in One True Theory, and that it will only be attained when a single point of view has been established beyond all doubt as the one and only correct account of the world. To this end he has explored at some length the techniques whereby the adherents of a particular theory attempt to win the allegiance of others to it, and to preserve it more or less indefinitely. More specifically, in his early work he analyzed the way in which philosophically-minded scientists attempted to limit the range of possible theories in the microphysical domain to one, viz. to the so-called Copengagen Interpretation of the quantum theory. It was in the context of Einstein's, Schrodinger's and Bohm's opposition to their dogmatic line of argument that Feyerabend initially developed some of his most important philosophical ideas.

On the basis of his own investigations into quantum theory Feyerabend satisfied himself that the physical arguments which underpinned Bohr's Copenhagen Interpretation were sound and genuine. However, he also came to the conclusion that

> despite this plausibility none of the arguments is powerful enough to guarantee the absolute validity of these ideas and to justify the demand that the theories of the microscopic domain will forever have to conform to a

certain pattern. Such restrictions are possible only if certain *philosophical* ideas are used as well . . . [which] are neither correct nor reasonable.[10]

Here there was a quite marked difference between the views of Einstein and of Bohr. Einstein asserted that all physical theories were fallible and subject to replacement. They were free mental creations in the construction of which the scientist 'must appear to the systematic epistemologist as a type of unscrupulous opportunist'.[11] Bohr and his collaborators, by contrast, believed that it was essential to found physical theories on a solid and incontrovertible empirical base. For them this base served as an external control on speculative thought, and restricted the scope of permissible scientific theorizing. By adopting this stance they fell prey to what Feyerabend calls the 'inductivist prejudice'. They regarded facts to be an unalterable basis for theorizing, and believed that theories which are produced on that basis are uniquely determined by the facts which support them. These views emerge in a variety of guises in the arguments for the exclusiveness of the Copenhagen Interpretation.

Two restrictions on innovation that Feyerabend finds particularly objectionable are the condition of meaning invariance and the consistency condition. Both are intended to ensure that there is an overlap between an established point of view and new theories that are put forward. By invoking them, Bohr and others aimed to enforce a measure of continuity in scientific growth and to preserve empirically successful theories in perpetuity.

The conditions of meaning invariance demands that there be *conceptual* continuity between an established view and a novel one. In Bohr's case it was invoked to ensure overlap between new physical theories and classical physics. Classical physics, said Bohr, was a universal conceptual scheme, which was firmly based on observation, which could accommodate all conceivable facts, and which had saturated every aspect of our interaction with the natural world. Therefore, as Feyerabend puts it, he concluded that it was impossible for us 'to conceive any concepts which fall outside the classical scheme'.[12] Feyerabend quotes Bohr and Heisenberg, respectively, as saying that '"however far the new phenomena" found on the microlevel "transcend the scope of classical physical *explanation*, the account of all evidence must be expressed in classical terms'",[13] and 'Any experiment in physics, whether it refers to the phenomena of daily life or to atomic events, is to be described in terms of classical physics . . . . *We cannot and should not replace these concepts by any others*'.[14]

Bohr and Heisenberg, then, invoke philosophically inspired arguments to preserve indefinitely their interpretation of the quantum theory. According to Feyerabend, their attempts to do so have received indirect support and legitimation from professional philosphers, particularly logical empiricists. Hempel's theory of explanation as well as Nagel's theory of reduction also assert, and demand, that there is and should be an essential continuity between any newly introduced theory and an empirically adequate alternative which already enjoys the support of scientists. To enforce an overlap between the two, both philosophers invoke the condition of meaning invariance and a related consistency condition. The latter demands that any new theory that is introduced to account for the same phenomena as an established theory must contain, or at least be consistent with, the theory which is already there. Consistency is achieved by specifying that a relationship of strict logical deducibility must hold between an existing theory and a novel one. For Nagel one way in which scientific knowledge grows is by our reducing a 'secondary' science to a more fundamental 'primary' one, and 'the objective of the reduction is to show that the laws, or the general principles of the secondary science, are simply logical consequences of the assumptions of the primary science'.[15] In similar vein, Hempel and Oppenheim demand that progress be achieved by making the transition from an existing theory to a wider one, which explains and preserves its successes in a particular domain: 'The explanandum [i.e., the existing theory] must be a logical consequence of the explanans; in other words, the explanandum must be logically deducible from the information contained in the explanans, for otherwise the explanans would not constitute adequate grounds for the explanation'.[16]

As I pointed out above, Feyerabend attributes these monistic tendencies in scientists and philosophers to the soporific effects of partial empirical success and, more especially, to the view that empirical theories are uniquely determined by the facts that support them. Progress is seen as being steadily cumulative, and as involving modifications and extensions to an edifice erected on a firm and irrefutable foundation. Questions of theory acceptance and rejection are discussed within the framework of a model which considers the relationship between a single theory and 'a never examined, mystical, and stable entity, "experience"'.[17] Accordingly, a discussion of alternatives to the accepted view is either deemed an irrelevant luxury or is postponed until the sheer weight of anomaly makes it glaringly obvious that there is something wrong with the prevailing orthodoxy. Until such time Kuhnian 'normal' science prevails — 'the monomaniac concern with only one single point of view'.[18] And even when the reigning paradigm collapses, the consistency condition and the condition of meaning invariance work together to enforce continuity between the new and the old, preserving the latter even while it is superseded by a richer and deeper conceptual scheme.

Feyerabend's assault on this position is informed by two kinds of argument. Firstly, he insists that it does not reflect actual scientific practice. A study of history reveals that the proliferation of alternatives to the dominant view goes on all the time in science. What's more, it is found that scientific advance has only occurred because theories were introduced which were inconsistent with those already there (e.g. Newton's dynamical laws are inconsistent with those formulated by Galileo and by Kepler), and because a prevailing observation language and its associated ontology was replaced by another (see the introductory comments to this chapter). In other words, science as we know it has been possible because the restrictions on proliferation imposed by monomaniacs and the proponents of continuity were violated in the past.

The second kind of objection which Feyerabend brings against theoretical monism is that it is unreasonable and violates the basic precepts of an empiricist methodology — which its adherents allegedly aspire to. Here his basic argument is that

one and the same set of observational data is compatible with very different and mutually inconsistent theories. This is possible for two reasons: first, because theories, which are universal, always go beyond any set of observations that might be available at any particular time; second, because the truth of an observation statement can always be asserted within a certain margin of error only.[19]

Either way, the empirical success of a theory does not guarantee that it is the One True Theory. In the first case, the fact that it has not yet been found wanting may simply indicate that we have failed to devise sufficiently strenuous tests for it. And, in the second case, those 'successful' tests that we have made may also provide empirical support for an entirely different theory whose observational consequences in the domain of overlap are not measurably different from those of the one which we favour.[21]

If this be granted, the restrictive effects on growth of the consistency condition are immediately apparent. For let us assume that a new theory is introduced which shares all the observational successes of the existing theory (within the margin of error) but which is inconsistent with it outside the zone of overlap. Then the consistency condition would eliminate the new theory, not because it is empirically inadequate, but simply because it contradicts prevailing orthodoxy. In effect, then, the untested part of the view at the centre of attention is being taken as establishing its superiority. This consideration works in its favour and not in favour of the more recent theory simply because the old and familiar view was there first, and not necessarily because of any inherent advantages that it may have. This is hardly the kind of consideration which should sway a good empiricist!

There is another way in which the consistency condition subverts empiricist methodology. Let us accept that a good empiricist aims to expose a theory to as wide a variety of tests as possible. Then Feyerabend points out that this can often only be done with the help of alternatives to the view which is being tested. To take but one example: the bending of light rays by a heavy body like the Sun, which provided empirical support for Einstein's theory, was incompatible with Newtonian physics. Yet Eddington's observations of the phenomenon during a solar eclipse would never have been made but for the Einsteinian alternative. In other words, it is a basic tenet of empiricism that the empirical content of our knowledge be increased as much as possible. The consistency condition violates this requirement by eliminating alternatives which can generate potentially refuting facts. This suggests, says Feyerabend, that 'the methodological unit to which we must refer when discussing questions of test and empirical content is constituted by a *whole set of partly overlapping, factually adequate, but mutually inconsistent theories*'.[21]

In the light of the above it is clear that empiricists who insist that proliferation should only begin once an established theory starts to break down put the cart before the horse. For alternatives that are inconsistent with the view at the centre of attention are necessary to expose its empirical inadequacy. In fact, the absence of any major difficulties may actually signify a *decrease* in its empirical content. It may simply mean that alternatives to the prevailing orthodoxy have not been seriously considered. Under such circumstances the theory is 'successful', not because it is true and corresponds with nature, but because facts incompatible with it have not been allowed to come to light. Its success, as Feyerabend likes to say, is entirely man-made.

To be a good empiricist, then, one must adopt the 'principle of proliferation: Invent, and elaborate theories which are inconsistent with the accepted point of view, even if the latter should happen to be highly confirmed and generally accepted'.[22] The function of such concrete alternatives is this

They provide means of criticizing the accepted theory in a manner which goes *beyond* the criticism provided by a comparison of that theory 'with the facts': however closely a theory seems to reflect the facts, however universal its use, and however necessary its existence seems to be to those speaking the corresponding idiom, its factual adequacy can be asserted only *after* it has been confronted with alternatives *whose invention and detailed development must therefore precede any final assertion of practical success and factual adequacy*.[23]

Methodological pluralism is imperative if we want to increase the empirical content of our theories. It is also psychologically important, for 'it allows scientists of different temperament to follow different inclinations'.[24] Unswerving allegiance to a single theory cripples our imaginative and innovative faculties, and inculcates dogmatic and conformist habits of thinking. A method that encourages variety is also the only method that is compatible with the development of well rounded individuals who are not cramped within the confines of a single conceptual framework. As for Hempel's and Nagel's consistency condition, which delimits variety, it contains, says Feyerabend, 'a theological element which lies, of course, in the worship of "facts" so characteristic of nearly all empiricism'.[25] Bemused by the partial empirical success of a prevailing orthodoxy, these logical empiricists, and the defenders of the Copenhagen Interpretation, have become the new enemies of progress in science. Being theoretical monists, they are as dogmatic and intolerant, says Feyerabend, as the 'school philosophers' of old.

In the arguments discussed above Feyerabend has isolated three levels of intellectual activity which mesh with one another to preserve Bohr's microphysics. Firstly, there is the Copenhagen Interpretation itself. Feyerabend regards this to be a sound and ingenious scientific theory of the microscopic domain which, though empirically successful, is not the only possible or even necessarily the best account of it. Secondly there are the philosophical ideas which scientific adherents of that interpretation invoke in an attempt to bestow absolute validity on it. Dominant here is the 'inductivist prejudice' that facts uniquely determine the theories which encompass them. And thirdly, there are the 'professional' philosophers (particularly logical empiricists) who have reformulated and refined some of the scientist's ideas, thereby stamping them with the mark of respectability — and contributing to the spread of conformist thinking. These philosophers, says Feyerabend, have betrayed their heritage. Earlier positivists like Mach, who adopted a critical attitude towards experience, were actively engaged in reforming the sciences. Hempel and Nagel, by contrast, are content merely to mirror them from afar. Carnap too is so intimidated by the achievements of modern science that he sees philosophy as doing little more than 'explicating' or making more precise its conceptual apparatus. Contemporary empiricism is then an essentially conservative, intellectually impotent exercize, which leaves everything as it is or, if it changes it at all, does so only in the sense that it further entrenches prevailing orthodoxy.

This has not always been so. In the seventeenth and eighteenth centuries philosophy was an ally of scientists who, unlike the conformists of today, '*smashed* the harmonious cosmos of earlier ages . . . eliminated fruitless "knowledge", changed social relations, and slowly assembled the elements of a new type of knowledge that was both true and beneficial for man'.[26] At this time philosophy was a revolutionary weapon which had 'a quite decisive role in building the new science and in defending new theories against their well-entrenched predecessors'.[27] The philosophy of science, as Feyerabend likes to put it, is a subject with a great past. I have a strong suspicion that he sees himself as heir to that past.

This point is of more than passing interest. Feyerabend is often accused of being an uncommitted, socially irresponsible Californian parasite, and has been called a play-philosopher. However, the argument in this section reveals him to be concretely engaged in an ongoing struggle within the sciences, siding with Einstein and Bohm against Bohr and against continuity. As such he is far more committed and involved than most of his critics, who usually do little more than pass high-sounding judgements from the relative comfort of the spectator's enclosure.

4.2 Against Method

The discussion in the previous section reveals two important and related aspects of Feyerabend's brand of philosophy. In the first place, he is vigorously opposed to dogmatism and intellectual conformism. This is because he takes them to be antithetical to his prime value, which is 'the happiness and the full development of an individual human being'.[28] And secondly, as I have stressed, Feyerabend involves himself as a participant in particular struggles for intellectual freedom. As such he argues against *specific* moves made by his opponents in the context in question to restrict the scope of speculative thought. Once this is appreciated, we are more easily able to grasp the significance of his approach and, in particular, of his 'methodological anarchism'.

To illustrate the point I am making, consider again the strategy that Feyerabend adopts in his critique of Bohr and his collaborators. What Feyerabend does is to detect the specific philosophical assumptions and views which the defenders of the Copenhagen Interpretation appeal to to solidify their theory. He connects these with the formal and more rigorous doctrines espoused by Hempel and by Nagel, in particular.

These philosophers, he notes, claim to be empiricists. One of the tasks which they undertake is to lay down the conditions which new theories must meet if they are to be accepted into the body of science. But, Feyerabend argues, the consistency condition and the condition of meaning invariance, which both Hempel and Nagel impose, violate empiricist methodology. In other words, Feyerabend is arguing that *if* one wants to be a good empiricist (and Hempel and Nagel do) *then* one should encourage proliferation.

Given this strategy it is therefore mistaken to attribute to Feyerabend the view that to achieve progress we *must* proliferate alternatives to the view at the centre of attention. What he does say is that *empiricists who are committed to progress in their sense* should proliferate theories, or at least should not stop others attempting to do so. The principle of proliferation is not a 'methodological rule' and it does not treat situations *sub specie aeternitatis*, as Feyerabend puts it. It is not universal in the sense that come what may it must be slavishly adhered to. It is only 'universal' in the sense that whenever empiricists who are committed to progress (understood as involving an increase in the empirical content of their theories) espouse monism, the principle can be invoked to show that their position is unreasonable.

The same considerations apply to Feyerabend himself. As I have said, his central value is individual happiness and not, shall we say, increase in empirical content. To achieve this goal, he argues, we must proliferate so that people can freely choose a cognitive system that they find pleasing. If individual happiness is not one's goal, then there may be no need to proliferate at all. Combining this argument with the former, we see that what Feyerabend shows is that monism is a bad philosophy for empiricists and for those who are concerned with individual intellectual freedom. More than that he does not show, nor does he claim to show.

The significance which Feyerabend attaches to the principle of proliferation is precisely that which he attaches to any rule or standard when he says 'I do not object to rules, standards, arguments, I only object to rules, standards, arguments *of a certain kind*. I object to rules, standards, arguments which are *general*, and *independent of the situation in which they are applied*'.[29] If the growth of knowledge is to be achieved by following any rules or methodology at all, it will have to be a 'methodology' which is both flexible and undogmatic

> Each rule, each demand, that it contains is asserted only conditionally, like a rule of thumb, and it can be overthrown, or replaced by its opposite, as the result of an examination of concrete cases . . . . Also, looking back into history, we find that for every rule one might want to defend, there exist circumstances where progress was made by breaking the rule. All this means that methodology can at most offer a somewhat chaotic list of rules of thumb and that the only principle we can trust under all circumstances is that *anything goes*.[30]

This is the central tenet of Feyerabend's methodological anarchism. It is, in the first instance, a demand for *freedom vis à vis methodological rules*. And of course, as Feyerabend points out, it is not really a methodological principle at all. It is simply a crutch which can be used by those who crave the intellectual security provided by having a principle which holds 'under all circumstances'

> If you want universal standards, I say, if you cannot live without principles that hold independently of situation, shape of the world, exigencies of research, then I can give you such a principle. But it will be empty, useless and pretty ridiculous. It will be the principle 'anything goes'.[31]

One important sense in which this slogan can be interpreted is as a contribution to a new conception of rationality. Seen from this angle, it reflects Feyerabend's view that if knowledge is to grow we have to be more flexible, imaginative and innovative than conventional methodologists will allow. It is a consequence of being concretely engaged in a historical process as a *participant*, and not as a spectator. On the basis of his own involvement with contemporary science, and of his study of Galileo's achievement, Feyerabend has come to the conclusion that, loosely speaking, reason differentiates: that a theory of rationality which attempts to treat concrete situations *sub specie aeternitatis* is bound to be inadequate. It simply fails to come to grips with the subtle complexities and nuances of the processes of change. As a theory of rationality, the slogan 'anything goes' amounts to a refusal to legislate in advance for each and every situation. As a theory of history it is an injunction directed at Popper, Lakatos and their admirers. It instructs them to analyze historical events *in their specificity*, and not to try to reconstruct them by fleshing out a skeletal structure of universal rules for 'rational' behaviour.

'Anything goes' liberates the participants in concrete situations from the need to behave in a rule-governed and predictable way if they wish to be rational. However, the freedom is bought at a price. The methodologies which Feyerabend criticizes sought to lay down universal criteria for theory choice and in doing so stressed the importance of empirical considerations as factors shaping ('rational') theory rejection and acceptance. To my mind Feyerabend is rightly sceptical about the value of a project of this kind. That granted, the onus is on him (and me!) to provide an alternative account of the pressures which steer historical agents across a revolutionary divide, and inform their commitment to a new theory. In the remainder of this chapter I shall explore the alternative which he develops through an analysis of the Copernican Revolution. I shall argue that that alternative is seriously flawed. For Feyerabend shares with Popper and with Lakatos the view that the process whereby new theories are produced is more or less arbitrary and rationally unintelligible — a matter for the empirical psychology of discovery. That bias imposes a structure on Feyerabend's analysis of the Scientific Revolution. By virtue of it he produces a grossly one-sided account of this event, an account which exaggerates the importance of urges and passions and psychological manipulation, at the expense of argument, of truth, of falsehood and of reason. The ultimate outcome is Dadaism, a trap into which Feyerabend falls partly because he has not yet freed himself from the paralysing grip of a rigid distinction between the context of discovery and the context of justification.

Having said that, a word of warning is necessary concerning Feyerabend's Dadaism. Feyerabend frequently refers to both the Ptolemaic and Copernican systems in Popperian language; he calls them refuted when he could describe them as being in empirical difficulty, for example. This suggests to me that at times he is less concerned with developing his own position on Galileo's achievement than with working out its implications for a Popperian. After all, he warns the reader against assuming that what he writes expresses his 'deep convictions'. Accordingly, even if I succeed in showing how Dadaism emerges out of Feyerabend's analysis of Galileo's behaviour, it does not follow that Feyerabend is a Dadaist. All that I will have shown is how someone who accepts a certain presuppositional framework and works through its implications in a Feyerabendian way can land up espousing Dadaism. This must always be borne in mind when I attribute views to Feyerabend and when I criticize them. As a matter of fact I neither know whether he is a Dadaist nor do I care. It is irrelevant for my purposes whether he adopts positions for the sake of argument or from conviction. I suspect that it is a complex mixture of both — as is only to be expected from one who is in a state of transition himself.

4.3 Feyerabend's Theory of Transition: The Challenge to Sense Experience

In forging a new conception of nature, the founders of modern science were forced to engage with, and to disentangle themselves from, a monolithic cosmology whose effects on consciousness were far-reaching. According to Feyerabend

> the only conceptual apparatus then available was the Aristotelian theory of change with its opposition of actual and potential properties, form and matter, the four causes, and the like. This conceptual apparatus was much more general and universal than the physical theories of today as it contained a general theory of change, spatio-temporal and otherwise. It also seems to be closer to everyday thinking and was therefore more firmly entrenched than any succeeding physical theory, classical physics included.[32]

Yet Copernicus and Galileo succeeded in breaking with the familiar, commonsensical views of the day. Sharing a commitment to a heliostatic universe, they developed the astronomical and dynamical theories demanded by their postulate that the Earth moved, and paved the way for classical mechanics. How did they manage to escape from the constraints imposed by a dominant and comprehensive theory like Aristotelianism? For Feyerabend, a crucial part of the answer lies in the attitude which they adopted towards sense experience.

It is easy for us to underestimate the difficulty which Copernicus and Galileo had in challenging the evidence of their senses, arguing that the Earth in fact moved. For it is now generally conceded that the *weakness* of Ptolemaic astronomy, and of Aristotelian cosmology in general, lay in their over-reliance on sense evidence. Ptolemy has been chastized for placing 'too great a reliance upon naïve observation . . . . The difficulty was that Ptolemy was too empirical'.[33] And the Aristotelians have been condemned for lacking 'a healthy distrust of the testimony of their own and others' senses'.[34] But these judgements are made with the wisdom of hindsight. For the ancients the view that the Earth was stationary at the centre of the universe and that the Sun moved daily around it seemed to be confirmed by the most rudimentary observations. And when, in 1543, Copernicus argued that the Earth rotates daily on its own axis, and moves annually around a stationary Sun, he was assailed on two fronts. Not only did Luther pour scorn on this 'upstart astrologer' who put forward views so plainly at variance with divinely revealed truths. One of his close associates, Melanchthon, also pointed out that 'The eyes are witnesses that the heavens revolve in the space of twenty-four hours'. Publicly to assert the contrary, he said, reveals a 'want of honesty and decency' ![35] It is not surprising that Copernicus began his preface to *De Revolutionibus* by expressing the fear that he would be 'hissed off the stage' when people realized that he had assumed that the Earth moved. As Galileo admiringly put it, both he and Aristarchus had 'through sheer force of intellect done such violence to their own senses as to prefer what reason told them over that which sensible experience plainly showed them to the contrary'.[36]

Of course it was not only astronomical phenomena which suggested that the Earth was stationary at the centre of the universe. Terrestrial observations also counted heavily against the idea of a moving Earth. Ptolemy, for example, had argued that if the Earth rotated daily from west to east, flying birds and clouds should be seen to move westwards at great speed — but they are not. What's more a stone thrown vertically into the air should fall some distance away from the person who propels it. The fact that it falls at that person's feet plainly refutes the idea that the Earth is moving. To these arguments others were added in the sixteenth century. Typically, it was pointed out that a cannon ball dropped from a tower on a moving Earth should strike its surface at some distance from the tower's base — a conclusion easily refuted by experience.

One of Feyerabend's favourite quotations is that in which Galileo deals with this objection. He accepted that a heavy body released from the top of a tower falls perpendicularly to the Earth but refused to draw the 'obvious' conclusion. Against an opponent who raised this objection to the Copernican system, Galileo remarked

> I wish that this author would not put himself to such trouble trying to have us understand from our senses that this motion of falling bodies is simple straight motion and no other kind, nor get angry and complain because such a clear, obvious and manifest thing should be called into question. For in this way he hints at believing that to those who say such motion is not straight at all, but rather circular, it seems they see the stone move visibly in an arc, since he calls upon their senses rather than their reason to clarify the effect. This is not the case, Simplicio; for just as I . . . have never seen nor ever expect to see the rock fall any way but perpendicularly, just so do I believe that it appears to the eyes of everyone else. It is therefore better to put aside the appearance, on which we all agree, and to use the power of reason either to confirm its reality or to reveal its fallacy.[37]

Unlike some modern physicists, then, Copernicus and Galileo were not victims of the inductivist prejudice, which holds that observational facts are an unalterable basis for theorizing, and that theories are uniquely determined by them. On the contrary they proceeded 'counterinductively' with respect to the facts: they advanced and were prepared to defend theories that were contradicted by the most obvious observational data. And Feyerabend insists that they were right to do so. For science develops unevenly, and different parts of it can be out of phase with each other. A clash between a new theory and existing observations need not mean that the theory is defective; rather it may occur because the evidence is 'contaminated'. The founding of a new science, I would say, involves a re-evaluation of experience, a redrawing of the boundary between appearance and reality. The status of self-evident 'facts' — like the 'fact' that the Sun moves around the Earth — changes accordingly.

In Section 3.1 I quoted Popper as saying that the question of theory acceptance and rejection ought to be settled primarily by recourse to observation and experiment. In the light of the above, however, it is plain that too heavy a reliance on empirical considerations can be disastrous, particularly for the

FROM DISCONTINUITY TO DADAISM

adherents of a new theory. Such theories are not only likely to have little empirical evidence in their favour when they are first put forward. If Feyerabend is right, the evidence that counts against them is also quite likely to be defective. The proponents of a new theory are thus out on a limb as far as the 'facts' are concerned. Something other than empirical considerations is called for to instil in them the confidence which they need to cling to their revolutionary hypothesis, particularly in its early stages.

In both of the quotations given above, Galileo sees *reason* as providing the proponents of the heliostatic view with the support they required to challenge the evidence of their senses. It was reason, he said, which 'became mistress of their belief' in Copernicanism. Feyerabend, however, re-interprets Galileo's claim in a significant way. Describing the Copernican rupture with Aristotelianism, he writes

> Modern astronomy started with the attempt of Copernicus to adapt the old ideas of Philolaos to the needs of astronomical predictions. Philolaos was not a precise scientist, he was a muddleheaded Pythagorean, . . . and the consequences of his doctrine were called 'incredibly ridiculous' by a professional astronomer such as Ptolemy . . . . Even Galileo, who had the much improved Copernican version of Philolaos before him, says: 'There is no limit to my astonishment when I reflect that Aristarchus and Copernicus were able to make reason so conquer sense that, in defiance of the latter, the former became mistress of their belief' (*Dialogue*, 328). 'Sense' here refers to the experiences which Aristotle and others had used to show that the earth must be at rest. The 'reason' which Copernicus opposes to their arguments is the very mystical faith ('mystical' from the point of view of today's rationalists) in the fundamental character of circular motion.[38]

Galileo's claims about reason, then, are taken by Feyerabend to refer to what he calls the 'mystical' reason of a muddle-headed Pythagorean plus a mystical faith in circularity. On his view, the defence of the heliostatic system in the face of recalcitrant evidence was not guided by reason, but by non-rational considerations. By combining a Platonic love for circles with a faith in the viability of an absurd hypothesis advanced some 2000 years earlier, Copernicus had the courage to risk being 'hissed off the stage' by the supporters of the geostatic, geocentric cosmology. For Feyerabend then the commitment to a new view which is empirically inferior to an established rival is dictated primarily by non-rational considerations. By this it is meant that the commitment is shaped by factors other than logic and experience, factors which, on that account, are construed as being non-rational. In other words Feyerabend adopts, if only for the sake of argument, the 'liberal' conception of rationality (see Chapter 1). This conception delimits the field of rational theory choice such that one slides almost inexorably from recognizing the limits of observation as a factor guiding belief into the view that the adoption of a theory necessarily contains a large non-rational component.

It is helpful to discuss Feyerabend's views on the scope of the non-rational in the light of Popper's and Lakatos' attitude to it. Like these two philosophers, Feyerabend sees the emergence of new ideas as being a more or less arbitrary event of the kind usually associated with the so-called context of discovery. Since he elaborates quite extensively on this it is worth pausing momentarily to look at what he says about the genesis of new ideas.

Feyerabend's general conception of how new world views arise reflects that desire to escape from the control exerted by 'facts' to which I alluded earlier. Proceedingly counterinductively, scientists must first bracket the empirical evidence which speaks in favour of an established view. According to Feyerabend, the freedom which they exercize in doing so is both methodologically and psychologically important. The latter is valued because 'it allows scientists of different temperament to follow their different inclinations, and thereby gives them satisfaction which goes beyond the satisfaction derived from the exclusive consideration of facts'.[39] The theories which emerge are not dictated by the facts, although they will enjoy a certain amount of empirical support. Instead they will reflect accidental features of the tradition in which the scientists work, as well as their personal idiosyncracies and prejudices. They will depend on elements which are rooted, not in facts, but in the mind of the theoretician and which are therefore 'subjective'.

Feyerabend suggests that the realms of fantasy and of imagination are valuable sources of material for these creative acts of theory innovation. It is they that have evaded the stranglehold on our consciousness of the prevailing orthodoxy. We need to turn to views which were educated out of us from infancy, when they were dismissed as absurd and irrational, and repressed accordingly. Such views 'turn up again, but in different shapes. They reappear as dreams, fairy-tales, literature, phantasy — that is, they are now all assembled under the heading of "fiction". But it is just ideas like these which we need in order to examine what the popular point of view tells us is "reality"'.[40] To them we can add those archaic and apparently ridiculous ideas which have lain dormant, perhaps for centuries, but which were superseded long before they had an opportunity to show their mettle. Together these several domains provide a rich source of raw material with which to build a new conceptual scheme which differs fundamentally from that at the centre of attention.

We see then that for Feyerabend the 'invention' (as he calls it) of alternatives to an established orthodoxy is not guided by reason. It is 'a contingent event — dependent on intelligence, relaxation and good luck'.[41] Both Copernicus and Galileo 'had to make a guess and in making this guess they could only follow their inclinations'.[42] Thus like Popper and Lakatos, Feyerabend locates the arbitrary and the contingent at the very core of the historical process. With them he holds that new theories emerge in a more or less random manner, and that the process whereby new knowledges are produced is not rationally intelligible.

This is an attractive position for many, not least of all because it sees scientific progress as demanding inputs from creative individuals who ought therefore to be left free to play with seemingly outlandish ideas. But it has risks, particularly for those who insist that, 'subjective' inputs notwithstanding, the history of science reveals the workings of what Hall calls the conscious, reasoning mind. For precisely how is the creative genius to be brought under the sway of reason? As I noted in previous chapters, Popper and Lakatos 'resolve' this dilemma by methodological fiat. Popper preserves the rationality of science by enforcing an overlap between novel conjectures and well established orthodoxy. Tradition as embodied in the scientific community imposes itself on the speculative thinker so as to ensure the 'objectivity' of science. Lakatos simply asserts that the nth version of one research programme will be 'blatantly, dramatically inconsistent' with the mth version of a rival (see Ref. 34, Section 3.5). He thereby conveniently sidesteps the difficulties precipitated by his philosophically-inspired refusal to consider seriously the historical processes whereby new theories or conceptual schemes are produced.

It is here that Feyerabend differs significantly from Popper and Lakatos. For he does not merely hold that new theories *emerge* in a more or less random manner. He also claims that the *initial choice* between a novel conjecture and an established view, say, is not guided by reason either. Thus he approvingly quotes Fichte as saying (in a somewhat different context) that the choice between comprehensive theories cannot be compelled by logic and experiment. Rather, it 'depends on the freedom of thought entirely. It is therefore determined in an arbitrary manner and, as an arbitrary decision must have a reason nevertheless, by our *inclination* and our *interest*'.[43]

When we remember Feyerabend's hostility to imposing methodological rules on practising scientists we can appreciate why, having accepted that theories emerge more or less randomly, he refuses to adopt Popper's and Lakatos' strategies for saving the rationality and the continuity of science. Anyway, as we shall see, he also holds that rational standards themselves change during revolutionary transitions. Galileo's remarks notwithstanding, he insists that reason cannot steer scientists across a discontinuous gap. Thus it is not surprising that Feyerabend should extend the scope of the arbitrary and the non-rational beyond theory emergence to theory choice. He would willingly agree with Kuhn that 'To astronomers the initial choice between Copernicus' system and Ptolemy's could only be a matter of tast . . . '.[44] It was only because Copernicus and Galileo followed their inclinations and ignored the dictates of reason that they were able to overlook the empirical inadequacies of their new view, and had the confidence to fight for the heliostatic scheme of the universe.

4.4 *Creating a Space for Copernicanism*

One of the most notable features of Feyerabend's theory of rupture is the extent to which he sees discontinuous changes as involving an historical agent's *disengagement* from the present. The emphasis which he shares with Popper and Lakatos on the value of individual creativity, of imagination, of intuition, and so forth, contains within it the seeds of this escapism. But whereas Popper and Lakatos attempt to circumscribe the freedom of thought, Feyerabend will not do so. He does not merely encourage people to step outside the present, to journey back to the past or to delve into a dream-world. He also suggests that if they find these alternative worlds attractive they should inhabit them, extending their boundaries while enriching themselves.

In Chapter 1 I quoted Clavelin as saying that Aristotelianism was 'the conceptual universe *from which* (and also *in opposition to which*) classical mechanics was founded' (Ref. 53, my emphases). Feyerabend's theory of discontinuous transition is quite different from this. By and large, he does not see new theories as emerging out of, and developing over and against a prevailing orthodoxy. Rather, they appear *alongside* it, and are propelled forward less by differentiating themselves from the view at the centre of attention than by what Feyerabend calls a 'vague urge' or a 'passion'.[45] For Feyerabend growth involves turning one's back on a world which one finds suffocating and restrictive; it is not achieved by simultaneously confronting that world and critically transforming it, perhaps with a view to eliminating it.

White attributes a similar view to Foucault. Discussing his *The Order of Things* he claims that Foucault's 'epistemic domains', which are similar to Kuhnian paradigms

> simply appear alongside one another — catastrophically, as it were, without rhyme or reason. Thus, the appearance of a new 'human science' does not represent a 'revolution' in thought or consciousness. A new science of life, wealth, or language does not rise up against its predecessors; it simply crystallizes alongside of it, filling up the 'space' left by the 'discourse' of earlier sciences.[46]

This perspective — a perspective from which Copernicanism is seen as 'crystallizing alongside' Aristotelian thought, rather than in opposition to it — profoundly influences Feyerabend's subsequent historical analyses. Consider for example his 'fairy-tale' on the rise of modern astronomy (Ref. 38). It is noteworthy that this account makes no reference to the state of astronomical theory at the time or, more to the point, to Copernicus' assessment of it. All that Feyerabend says is that certain astronomical predictions were required and that there were observational data which counted against the movement of the Earth. If that is all that was at stake though, why did Copernicus put forward his new system ? After all Ptolemaic astronomy was immensely flexible and there were dynamical objections to treating the Earth as a planet — which Copernicus made only a half-hearted attempt to deal with. Did Copernicus set the Earth in motion primarily because he was inspired by a 'typical metaphysical urge'[47] for harmony and simplicity, an urge which he was able to indulge because of the freedom granted to theorizing by the indeterminateness of facts ? Feyerabend suggests that he did; that inspired by the ramblings of a 'madman' and moved by a Platonic love for circles Copernicus ignored the arguments against the whirling of the Earth and proposed his heliostatic theory. What role, if any, Copernicus' views on traditional astronomy had to play in this process Feyerabend does not say.

Feyerabend's analysis of Galileo's achievement reveals similar tendencies. Significantly, he repeatedly quotes a remark made by Galileo in *The Assayer* that both the Ptolemaic and the Copernican systems are 'surely false'.[48] This immediately places the Two Chief World Systems on the same epistemological footing, as it were. How did Galileo proceed in such circumstances ? Obviously his commitment to Copernicanism cannot be based simply on the belief that Aristotelianism is *mistaken*; nor can his defence of it be seen as involving a search for truth, for he allegedly believes that Copernicanism is false anyway. As we have seen, Feyerabend claims that one of Galileo's prime motives for developing the heliostatic theory was that he found it to be attractive and interesting. Overlooking the fact that his preferred cosmology was refuted, Galileo used the freedom which we have *vis à vis* the observational base to follow his inclinations and to develop an alternative to the Aristotelian system which he felt to be pleasing.

Of course in developing his alternative Galileo had to take account of empirical considerations. What I now want to do is to discuss briefly two of his arguments in defence of Copernicanism which Feyerabend explores in some detail. These are the tower argument against, and the telescopic evidence for, the movement of the Earth. As we shall see, the way in which Feyerabend interprets Galileo's treatment of observational material is particularly instructive. On the one hand it directly reflects those tendencies in his theory of transition which I have just discussed. On the other hand, it leads naturally into his espousal of Dadaism.

The so-called tower argument, it will be remembered, asserted that if the Earth moves a heavy body released from the top of a tower should strike the ground some distance away from its base. Since heavy bodies are not seen to behave in this way it was obvious, so the argument ran, that the Earth was stationary. Feyerabend argues that dynamical objections of this kind to terrestrial motion are particularly tenacious because they are based on what he calls 'natural interpretations'.

Natural interpretations are 'mental operations which follow so closely upon the senses'[49] that it is immensely difficult if not impossible to separate the associated beliefs from the sensory evidence which seems to speak unambiguously in their favour. The belief that the Earth is stationary is of this kind; the sensory evidence which supports it is overwhelming and is 'spontaneously' connected with it. (Feminists argue that the belief that socially constructed differences between the sexes are, in fact, biologically determined is similarly induced.[50]) Deeply entrenched in the common sense of the seventeenth century, the idea that the Earth did not move rested on a prevalent paradigm of motion which took it to be 'operative'. In other words it took as its model of motion a situation in which to move was to undergo a perceptible displacement with respect to an observer.

In dealing with the tower argument Galileo was therefore immediately at a disadvantage. To defuse this argument he introduced a relative concept of motion. The stone grazes the tower, he says, because although the Earth is in motion

> with respect to the earth, the tower, and ourselves, all of which all keep moving with the diurnal motion along with the stone, the diurnal movement is as if it did not exist; it remains insensible, imperceptible, and without any effect whatever. All that remains observable is the motion which we lack, and that is the grazing drop to the base of the tower.[51]

If this argument was to be effective, says Feyerabend, Galileo needed people to appreciate that among things which shared it in common, motion was non-operative (i.e., that it did not produce observable effects). However, Galileo himself remarked in the *Dialogue* that his opponents 'felt a great repugnance towards recognizing this nonoperative quality of motion among the things which share it in common'.[52] How then was he to proceed ? How was Galileo to overcome the resistance of his opponents to the relativity principle ?

Feyerabend suggests that Galileo used the Platonic technique of *anamnesis*. He reminded the Aristotelians that they were already familiar with the workings of the relativity principle, which they experienced directly when on a boat or a moving carriage. He then persuaded them to apply it to the Earth as well, suggesting that the readiness to do so was always in them 'although it took some effort to make it conscious'.[53] By using these psychological tricks the required conversion is achieved. Summarizing, Feyerabend claims that the change that took place was one in which 'an inadequate view, the Copernican theory, is supported by another inadequate view, the idea of the non-operative character of shared motion, and both theories gain strength and give support to each other in the process'.[54] As a result, a space is created alongside Aristotelianism in which the heliostatic system can survive and grow.

Feyerabend's analysis of Galileo's appeals to telescopic observations in favour of Copernicanism follows a similar trajectory. It is often said that the telescope, first used by Galileo in the summer of 1609, provided independent evidence for the heliostatic system. Kuhn, for example, alleges that 'In Galileo's hands the telescope disclosed countless evidences for Copernicanism'.[55] Feyerabend is more sceptical. He points out that although the telescope received due acclaim for its success when used to view objects on the Earth, many people had great difficulty in interpreting the images obtained when it was trained on the heavens. For example, on one disastrous evening in April 1610 Galileo took his instrument to the house of an opponent to demonstrate its powers. It was reported that no one present was convinced that Jupiter had satellites, as Galileo had claimed in *The Starry Messenger*, and the assembled dignitaries could not but conclude that they were probably fictitious. What's more Galileo's own recorded observations were hopelessly inaccurate. His drawings of the Moon show only one sharply defined crater. Its diameter is more than twice that of the largest craters in this part of the Moon, as measured by modern methods, and its size is such as to place it well above the threshold of naked eye observation — by means of which it is nevertheless not detected. In short, as a pupil of Kepler's was to say of the telescope; 'Below it works wonderfully; in the heavens it deceives one . . .'[56]

There is some doubt, then, as to whether Galileo's telescopes were a 'superior and better sense' than the unaided eye for viewing celestial phenomena. Only a theory of optics could resolve the ambiguities in its results. However, Feyerabend insists that Galileo did not know the optical theories of the time and that even if he had they would not have helped him very much because they were not sufficiently sophisticated to build a new telescope.

On the other hand there were phenomena which supported Copernicanism when viewed through a telescope but which undermined it when viewed with the naked eye. For example, Copernicus predicted that as it traverses its orbit Venus' brightness should vary by about a factor of 40. With the unaided eye, however, changes in its brightness were barely perceptible. With the telescope they were easily detected. By cutting out spurious rays, which bedecked the body of the planet when it was close to the Sun, the telescope gave a better indication of its true brightness than that gained with the naked eye — or at least so Galileo thought. Feyerabend claims that we are still unable satisfactorily to explain the disappearance of irradiation in the telescope. Be that as it may, the results obtained confirmed Copernicus' predictions.

Which then are we to trust ? The telescope or the unaided eye ? Galileo opts for the former, not because he has an adequate understanding of how the telescope works, says Feyerabend, but because the variations in brightness as seen through the telescope support Copernicanism. Galileo *presupposes* the heliostatic system and uses it to select between conflicting observational reports which he has no other means of assessing. As Feyerabend puts it, the telescope reveals some rather puzzling (with respect to naked eye observations) changes in planetary brightness

But the change is in harmony with the predictions of Copernicus. *It is this harmony rather than any deep understanding of cosmology and of optics which for Galileo proves Copernicus and the veracity of the telescope* in terrestrial *as well as* celestial matters. And it is this harmony on which he builds an entirely new view of the universe.[57]

The general conclusion that Feyerabend draws from his analysis of the tower argument and Galileo's use of telescopic evidence is that

> while the pre-Copernican astronomy *was in trouble* (was confronted by a series of refuting instances and implausibilities), the Copernican theory *was in even greater trouble* (was confronted by even more drastic refuting instances and implausibilities); but that being in harmony *with still further inadequate theories* it gained strength, and was retained, the refutations being made ineffective by *ad hoc* hypotheses and clever techniques of persuasion.[58]

And this only goes to show that science as we know it could never have arisen if Galileo had not been a methodological anarchist. For, writes Feyerabend 'Inventing theories and contemplating them in a relaxed and "artistic" fashion, we often make moves that are forbidden by methodological rules. For example, we interpret the evidence so that it fits our fanciful ideas, we eliminate difficulties by *ad hoc* procedures, we push them aside, or we simply refuse to take them seriously'.[59]

Feyerabend's analysis of Galileo's procedure stresses that the latter used his methodological freedom to *defuse* objections that its critics levelled against Copernicanism. And indeed in this respect the analysis correctly reflects Galileo's intentions on the Second and Third Days of the *Dialogue*, from which Feyerabend draws so much of his historical material. During these two days Galileo frequently said that his aim was not to prove that Copernicanism was true. It was merely to show that, like its Ptolemaic rival, it was compatible with the empirical evidence, which was usually thought to count against it. As Galileo said at the start of the Fourth Day of the *Dialogue*, the preceding discussion had shown that 'all terrestrial events from which it is ordinarily held that the earth stands still and the sun and the fixed stars are moving would necessarily appear just the same to us if the earth moved and the others stood still'.[60] Galileo's aim, in other words, was not to oppose and to reject the Ptolemaic system at this point, but to protect Copernicanism from allegations that it was empirically defective. On these two Days at least, he sought to create a space for his preferred view *alongside, rather than in opposition to*, the prevailing orthodoxy.

By locating this feature of Galileo's procedure within his own philosophical framework, Feyerabend has produced a grossly one-sided account of Galileo's defence of the heliostatic system. Galileo's initial commitment to the theory was allegedly 'nothing but a blind faith'[61] in the viability of a cosmology which he found pleasing; the new hypothesis 'gained strength' through the harmony between it and further inadequate theories. 'Refutations' were made 'ineffective', difficulties were 'pushed aside'. On this basis, and by these procedures, 'an entirely new view of the universe' was developed. From this perspective to accept that Galileo did not produce independent evidence for Copernicanism is to see him as simply rationalizing his allegiance to that world view. What then is Galileo doing but showing us 'how easy it is to lead people by the nose in a rational way ?'[62] But 'inventing compelling reasons for unreasonable doctrines ?'[63] By attributing Galileo's commitment to Copernicanism to faith, to a vague urge, to inclination, and by appreciating that in much of the *Dialogue* Galileo was not concerned to establish the Copernican hypothesis, Feyerabend is well on the way to turning him into a Dadaist !

4.5 Relativizing Reason

We are now better able to appreciate the implications of Feyerabend's view (which I share) that science develops unevenly. Because it does, those who espouse a new, revolutionary theory take an enormous risk. The established tradition confronts them as a monolithic, successful, and generally accepted mode of cognition. Given its proven ability to absorb anomalies, its adherents are confident that it will be able to continue to cope with difficulties when they arise. On 'objective' grounds the rival for which a Galileo is prepared to fight is grossly inadequate by comparison. It is not only more limited in its scope. It is also 'born refuted'. Accordingly Feyerabend insists that only 'subjective' factors can instil in revolutionaries the determination that they need to stick to their guns in the face of adversity. It is inclination and passion which are the driving forces of history, not reason and argument.

In the teeth of opposition, the proponents of a new view set out to accumulate empirical evidence for it. As I showed in the previous section, Feyerabend believes that their prior commitment to their cosmology plays a crucial role in this process. The evidence adduced in its favour does not establish its truth. It is gleaned on the assumption that the theory is true. It is only long after people have probed the unknown drawing strength from 'subjective' factors that 'objective' judgements between empirically rich, 'ready, articulated theories' (Lakatos) are possible — and by then they will be irrelevant anyway. As Feyerabend puts it

I cannot believe . . . that a revolution such as the Copernican Revolution proceeded in the full consciousness of the ideas and methods, and with a full understanding of the instruments about (i.e., within the next 300 years) to be invented. In all these cases the element of *action* — unreasonable, nonsensical, mad, immoral action when seen from the point of view of a contemporary — is a necessary presupposition of whatever clarity one would like to *possess*, but can achieve only *after the event*, as the *result* of the actions performed.[64]

And against those philosophers who naïvely think that progress ought to be achieved by following a set of rules Feyerabend claims that, in bursting the bonds of prevailing orthodoxy, the scientist

acts without any explicit rules and by his actions *constitutes* rationality; otherwise science would never have arisen and scientific revolutions would never have occurred. We must realize that science is much more flexible and much more difficult than rationalists assume it to be. A scientist is an inventor not only of theories but also of facts, standards, forms of rationality, in a word, an inventor of entire forms of life and his work can be understood only by those who are able to comprehend the details of such inventions.[65]

Again, concerning the specific point of how new conceptions of reason are constituted during periods of discontinuous change, Feyerabend has this to say

Creation of a *thing*, and creation plus full understanding of a *correct idea* of the thing, *are very often parts of one and the same indivisible process* and cannot be separated without bringing the process to a stop. The process itself is not guided by a well-defined programme, and cannot be guided by such a programme, for it contains the conditions for the realization of all possible programmes. It is guided rather by a vague urge, by a 'passion' (Kierkegaard). The passion gives rise to specific behaviour which in turn creates the circumstances and the ideas necessary for analysing and explaining the process, for making it 'rational'.[66]

There are some interesting similarities between these ideas on discontinuous change and conclusions reached by Foucault, who approached the problem from a somewhat different direction. He too has suggested that during periods of radical transition our conception of what is rational emerges out of and in response to quite specific concrete practices; reason does not lead them, it lags them. He studied changing conceptions of unreason in the West from the sixteenth to the twentieth century. In a helpful discussion of his views White points out that Foucault concluded that embedded in the ways in which the 'insane' were treated during this period were conceptions of madness and of the rational that differed substantially from the conceptions developed by writers on these subjects at the time. By shifting the boundary between the rational and the insane these writers effectively *rationalized* the imposition of constraints on socially disruptive individuals and groups. Foucault, writes White

> claimed that the history of madness revealed no consistent progress in the theoretical conceptualization of it as an illness, that, on the contrary, the history of the treatment of the insane revealed a consistent tendency to project very general social preconceptions and anxieties into theoretical systems which justified the confinement of whatever social group or personality type appeared to threaten society during a particular period.[67]

Admittedly Feyerabend allows for a more dynamic interaction between 'theory' and 'practice' than this. For he suggests that the creative process goes hand in hand with an increasingly 'full understanding' of it. Yet he too is ultimately sceptical about the eventual role which conceptions of reason play in establishing a form of life. Consider Newton. He proposes a monistic methodology, says Feyerabend, and claims that he has obtained his optics and his celestial mechanics by rigorously applying it. In fact he violates every single rule that he proposes. But his methodological prescriptions, and his manner of presentation give the impression that he has proceeded rationally. Newton, says Feyerabend, 'Having convinced most of his contemporaries,

... creates additional support both for his science (it has been obtained in a methodologically sound way and must therefore be free from major mistakes) and for his methodology (it has led to perfect scientific results and must therefore be the correct method)'.[68] In this way a conception of rationality is generated which bears little or no relation to the actual practice of science. Enshrined in the 'objective' methods which supposedly account for its success, it is used to devalue alternative and opposing modes of cognition as non-rational.

Feyerabend levels a similar criticism at Lakatos. As we have seen, Lakatos has suggested that an adequate theory of rationality must be consistent with the basic appraisals of the scientific elite of the past two centuries. These are value judgements about specific achievements which are accepted by the majority of scientists. This proposal, Feyerabend stresses, *presupposes* that by and large the members of the scientific elite in question have behaved rationally. In so doing it overlooks the fact that claims made about the rationality of one's own cognitive system are little more than rationalizations for one's own preferences and prejudices. The conception of rationality which is extracted from the past behaviour of scientists thus simply reflects the professional ideology of the scientific establishment. It does not argue for the cognitive superiority of their position, and it cannot be used to critcize that position as being itself unreasonable.

Once it is conceded that conceptions of reason are historical products which are generated to protect the interests of a specific group, reason (conceived as a set of rules and standards for ensuring success) is itself relativized. It is open to different groups to explore the possibilities inherent in their theories or modes of cognition and, having achieved success, to formulate an appropriate conception of 'rational' behaviour which they think can account for it. What one group cannot do is to use its standards to evaluate as irrational the practices of the adherents of alternative systems of thought. Nor can it attempt to stop them on those grounds from pursuing ideas which they find attractive and useful. Only social and psychological pressures, dressed up in the trappings of 'reason', can do that.

Thus do we arrive at Dadaism. The anarchist or Dadaist, Feyerabend tells us, is like 'an undercover agent who plays the game of Reason in order to undercut the authority of Reason (Truth, Honesty, Justice, and so on)'.[69] This devaluation of reason by the Dadaist derives its intellectual justification from the line of argument exposed above. In the first place, Feyerabend has suggested that driven on by urges and passions the scientific revolutionaries of the sixteenth and seventeenth centuries committed themselves to a cosmology which was plainly inferior to the orthodoxy of the day. The reasons which they gave in defense of their position amounted to little more than rationalizations of their subjective wishes, and the role of argument in their struggle was defensive. They had found a different way of life, and they wanted to be left alone to live it as they thought fit. Subsequently the practices, which in their view were constitutive of that form of life, had the honorific title of 'the rational' bestowed upon them. This leads on to Feyerabend's second point, that there are no universal rules or standards of rationality. Accordingly, critics who condemn as irrational cognitive systems or modes of behaviour of which they do not approve are doing little more than indulge their prejudices. Having developed a theory of rationality which legitimates their own form of life they now wish to impose it on others in the name of reason and progress.

Dadaists see through this pretentious charade. They are disgusted by the self-righteous and smugness of the standard bearers of Reason and Truth, and take immense pleasure in leading rationalists by the nose by 'inventing compelling reasons for unreasonable doctrines'.[70] For them, humour is liberty. Those who play off one view against another with intensity and conviction are making asses of themselves. To grow what is needed

is less moralism, *less seriousness*, less concern for the truth, a vastly deflated 'professional conscience', a more playful attitude, conventionalization of 'a lack of deep feeling' . . . and a good deal of laziness — and we shall be able to have our cake: to have freedom of choice in practical as well as in intellectual matters — and to eat it: to have this freedom without too much mental and emotional strain.[71]

Criticism ought to be an entertaining interchange between different groups of people, each of which respects the others' right to their own views. It is a battle of wits played according to no strict rules, which one enters and leaves as the fancy takes one. Nothing hinges on the outcome, for all participants realize that they are doing no more than rationalizing theories that appeal to them, mischievously defending with ingenious arguments the most outlandish ideas. But then, as Dadaists, they know that 'after all, reason *is* the slave of passions'.[72]

4.6 *Feyerabend and Philosophy*

When discussing the Copenhagen Interpretation of the quantum theory I pointed out that Feyerabend had isolated three levels of intellectual activity which worked together to build and to consolidate a new scientific theory. These were the practice of science itself, the scientist's philosophical reflections on that practice, and the further refinement of those reflections by professional philosophers. We can now see that this is part of a far more general conception of the changing relationships between philosophy and the sciences which is an integral component of Feyerabend's theory of discontinuity.

Feyerabend's theory of discontinuity, it will be remembered, starts from the claim that facts are not an unalterable basis for theorizing, and that no single theory is uniquely determined by them. This freedom granted to the theoretician *vis à vis* the observational base has two quite different kinds of consequences which can lead, in turn, to two different and contradictory roles for philosophy. On the one hand, it means that the empirical success of an existing theory does not guarantee its validity and need not intimidate scientists who wish to develop alternatives to a 'successful' orthodoxy. Philosophy can here be used to criticize the inductivisit prejudice of those who take partial empirical success as indicating absolute truth. On the other hand, by the selfsame argument, those who want to win and to cement people's allegiance to their particular point of view need something more than an appeal to its observational success in order to do so.

Philosophy now plays a quite different role. As they begin to assert themselves, the proponents of the new theory behave as if 'there is only one correct point of view; the correct philosophical method aims at proving its truth; and the correct psychological procedure aims at establishing unanimity as well as steadfastness in the pursuit of truth'.[73] Accordingly, with the passage of time, and if the propaganda for a new cognitive system is successful, a once subversive point of view will turn into a petrified dogma. And it will be helped on its way by a particular conception of reason which has been transformed from a revolutionary weapon, building an alternative to the prevailing orthodoxy, into a conservative doctrine mirroring and legitimating the latest stultifying ideology.

The overall effect is a general hardening of responses. Creative and imaginative individuals are forced to repress their innovative energies or are condemned to marginality. Alternatives to the prevailing orthodoxy seem to be absurd and irrational, the ramblings of madmen and incompetents. A dull conformism to the One True Theory or cognitive system is produced. Such unanimity, says Feyerabend, may be fitting for a church, for the victims of some ancient or modern myth, or for the spineless followers of a tyrant. Variety of opinion is essential both for objective knowledge and for individual fulfilment.

There is a clear line of development from Popper through Lakatos to Feyerabend concerning the role of reason in the growth of knowledge. All three grant to the non-rational a crucial role in the development of scientific thought, it being the source of new ideas. This is part of a conception of history which sees the historically disengaged, intellectually productive individual as one essential motor of progress. Popper immediately restricts the freedom thus gained in the so-called context of discovery by combining *modus tollens* with a set of methodological rules in the so-called context of justification. Together these specify what moves in the game of science are reasonable, emphasizing particularly the importance of eliminating empirically defective theories. Influenced by Kuhn's historical studies, Lakatos loosened the constraints in the context of justification. He found that it was extremely difficult to condemn as irrational scientists who refused to abandon a research programme, even though on his own standards that programme was degenerating. The best that he could do was to suggest that socio-psychological pressures should be used to remove it.

Emphasizing the importance of individual freedom, and developing a conception of reason which sees it as essentially serving group interests, Feyerabend pushed Lakatos' position to what was, after all, only its logical conclusion. There are no rational grounds for eliminating any cognitive system that people find helpful and attractive. The history of thought is a history of bodies of knowledge which spring up arbitrarily alongside one another and which would develop in parallel in perpetuity if 'external' pressures did not force them beneath the surface of consciousness, from which they may well up again at any time.

Dadaists recognize this. They realize that the triumph of reason is hollow, and that the conformity which it imposes is crippling. They insist that world views have been crushed in the name of reason which were as 'rational' in their own terms as any other cognitive system. That granted, they claim that the only humane and intellectually authentic slogan is *anything goes*, which means that, in practice, *everything stays*.

4.7 The Collapse of Oppositional Criticism

There is a simple theoretical distinction between alternative and oppositional, that is to say between someone who simply finds a different way to live and wishes to be left alone with it, and someone who finds a different way to live and wants to change the society in its light. This is usually the difference between individual and small-group solutions to social crisis and those solutions which properly belong to political and ultimately revolutionary practice. But it is often a very narrow line, in reality, between alternative and oppositional.[74]

There is a fundamental tension permeating Feyerabend's work. It is that between his appreciation of the importance of criticism in the light of alternatives, on the one hand, and his commitment to individual happiness and fulfilment on the other. He values the latter above all else, and his conception of the role and nature of criticism is adjusted accordingly. It is, as I mentioned above, essentially defensive: it seeks to create spaces for different points of view alongside one another. Out of this preference for the alternative at the expense of the oppositional Feyerabend ultimately extracts a position which is as cynical regarding the value of argument as it is politically conservative. That position is Dadaism, a position that tolerates everything, including the intolerable.

There is an intimate link between Feyerabend's rejection of

reason and truth, his preference for defensive rather than oppositional criticism, and his uncompromising demand for individual liberty. This is clear from his claim that knowledge grows through combining proliferation, which allows for the introduction of new theories and ideas, with tenacity, which allows for their articulation and development

> Proliferation means that there is no need to suppress even the most outlandish product of the human brain. *Everyone may follow his inclinations* and science, conceived as a critical enterprise, will profit from such an activity. Tenacity: this means that one is encouraged not just to follow one's inclinations, but to develop them further, to raise them, with the help of criticism (which involves a comparison with the existing alternatives) to a higher level of articulation *and thereby to raise their defence to a higher level of consciousness.*[75]

Feyerabend claims that these two methodological principles are compatible with individual liberty in that together they apply minimum pressure on us to deviate from our natural inclinations.

If proliferation means that *everyone* may follow their inclinations then it means that those who find a prevailing orthodoxy attractive, as well as those who like 'outlandish' alternatives may do so. That granted, the point of criticism is not to get people to relinquish a theory which they find pleasing. It is to help them to discover its strengths. The principle of tenacity embodies this idea. It justifies criticism on the grounds that it will enable the adherents of a particular point of view to raise their *defence* of it to a higher level of consciousness. In other words, the adherents of an established theory, as well as of a novel hypothesis, are being encouraged to tolerate criticism from different angles on the grounds that it is in their best interests to do so.

At least as far as the established theory is concerned, this is precisely what Feyerabend condemned Wittgensteinians for doing in one of his early papers. They encourage proliferation of alternatives, says Feyerabend

> not in order to arrive at a better theory through a criticism of the existing one, but rather *in order to get a better insight into the existing theory*. And getting a better insight into the existing theory (the existing 'language game') means revealing its hidden strengths, i.e. its capability to deal with the problems that have arisen and to remove the impression that a revision might be

needed. A Wittgensteinian uses alternatives with a *dogmatic*, or a *conservative*, purpose.[76]

Whether or not this is true of Wittgensteinians, the same conservative tendencies inform Feyerabend's recent attempts to combine criticism with proliferation.

In similar vein, Feyerabend sees the freedom that he demands *vis à vis* methodological rules as being used primarily to defend and to strengthen a particular point of view, rather than to eliminate an alternative. He is not always consistent on this score, however. Now and again he suggests that the adherents of a new hypothesis need to be flexible so as to destroy support for a dominant rival. How else, he says

> can we persuade people to follow our lead ? How can we lure them away from a well-defined, sophisticated and empirically successful system and make them transfer their allegiance to an unfinished and absurd hypothesis ? . . . . It is clear that allegiance to the new ideas will have to be brought about by means other than arguments. It will have to be brought about *by irrational means* such as propaganda, emotion, *ad hoc* hypotheses, and appeal to prejudices of all kinds.[77]

This aggression is necessary, because sometimes 'We need to be virtually bludgeoned into detachment from our daily lives, our habits and mental laziness, which conceal from us the strangeness of the world'.[78] However, no sooner has he said this than Feyerabend goes on to insist that, nevertheless, criticism is *'dangerous* unless we can show that a society enjoying criticism creates greater human happiness'.[79]

It is the pressure exerted by this consideration that ultimately prevails. Thus having just insisted that 'irrational' means (i.e., means other than argument) are needed to get people to switch their allegiance to a new theory, Feyerabend immediately goes on to say that such means are needed to *defend* that theory from the encroachments of a more powerful rival. The quotation given above (Ref. 77) continues that 'We need these "irrational" means in order to uphold what is nothing but a blind faith until we have found the auxiliary sciences, the facts, the arguments that turn the faith into sound "knowledge"'.[80] Once again Feyerabend is suggesting that the adherents of a new theory ought to leave the supporters of an established view to relish a system which they find pleasing, asking only that they be allowed to do the same in return. It is primarily for this purpose, for the purpose of parallel development, that Feyerabend insists on the value of methodological anarchism — and Dadaism.

Adopting the happiness and full development of an individual as a basic value, we need not only a methodology but also a set of institutions to ensure that no one is coerced more than is necessary to deviate from his or her natural inclinations. The institutional structure which Feyerabend advocates is posited on 'the principle of minimal interference with the life of individuals, or groups of individuals who have decided to pursue a common aim'.[81] This includes modern scienticists who will be allowed to continue providing us 'with pills, gas, electricity, atom bombs, frozen dinners and, occasionally, with a few interesting fairy-tales',[82] but who will not be allowed to impose their cosmology on the minds of the young at the expense of alternatives like the religious and the magical world views. Although Feyerabend claims that modern science is an oppressive, chauvinistic, big business, in the control of ants parading as experts, and that it has produced a hectic barbarism, his prime aim is not critically to transform it, but merely to limit its influence. The state is supposed to do this. The separation between the state and religion is to be supplemented by a separation between the state and science and, indeed, between the state and any ideology. Floating freely, as it were, above all ideologies, the state's role will be to ensure that no single subsociety — including science, the prevailing orthodoxy — imposes itself on any other.

4.8 *Down with Dadaism, Up with Discontinuity*

Throughout this chapter I have emphasized the importance of what Feyerabend calls the uneven development of science. By virtue of it the revolutionary project is an essentially risky one. The adherents of a new theory are confronted with immense difficulties, not the least of which is the empirical inadequacy of their views. In this chapter I have insisted that for Feyerabend they derive the courage they need to cling to their novel hypotheses from urges, passions, inclinations, faith, or some other basically non-rational factor. It has been my contention that by theorizing revolutionary commitment in this way Feyerabend slithers from discontinuity to Dadaism.

There is, however, something basically paradoxical about Feyerabend's extraction of Dadaism from his study of revolutionary change. For as I suggested in the opening remarks of this chapter, one cannot consistently be a Dadaist *and* a revolutionary. Dadaists who believe that 'anything goes' leave the establishment alone. They do not oppose it with a view to eliminating it and replacing it with their preferred form of life. Yet this is precisely what happens during revolutionary, discontinuous transitions.

Feyerabend's espousal of Dadaism amounts to a rejection of revolution, then. He adopts this stance because he has come to believe that discontinuous transitions are effected by coercion and force masquerading as reason. His arguments suggest that revolutionaries have done little more than beat their opponents into submission in the name of progress and of truth, all the while self-righteously rationalizing their commitment to their preferred cosmology. As far as Feyerabend is concerned, those who care about individual happiness and fulfilment had better recognize this. If this is one's prime value — and it is his — one had best become a Dadaist and forget about being a revolutionary.

It is plain that we are here dealing with an immensely complex issue which I cannot presume to resolve in a straightforward way. On the other hand, I have stressed that Feyerabend's emphasis on the non-rational as a motor of growth is one symptom of a basic flaw in his epistemology. Like Popper and Lakatos, he regards the emergence of new ideas to be a contingent and more or less random event. This philosophical prejudice profoundly affects his subsequent analysis of revolutions, as I have shown. Thus before capitulating to Feyerabend's position — which may ultimately be unavoidable — it is important that we look more closely at the circumstances under which new knowledges are produced. In so doing we may be able to place revolutionary commitment on a more rational basis than that of urges, passions, faith, and so forth.

In this connection it is instructive to consider, if only superficially, the sources of revolutionary energy identified and argued for by political theorists. This approach has the added advantage of suggesting interesting parallels between Feyerabend's epistemology and certain tendencies in political thought which may be worth investigating further on another occasion.

Consider first the similarities between Feyerabend and Marcuse, both of whom emphasize the importance of the repressed and the 'artistic' as sources of material for constructing alternative conceptions of reality. One of Marcuse's concerns is to retrieve from Marx's work a vision of what a socialist society would be like. What he finds distressing is that Marxist theoreticians have tended to overlook the fact that the young Marx, at least, thought that socialist society would be *qualitatively* different from all preceding societies in that labour would no longer be alienated. Marxists focussed on a more rational organization of the labour process with a view to expanding the productive forces 'and the result was some kind of frightening continuity between the developed capitalist societies and the image of a socialist society'.[83] Their failure to transcend the cognitive limits that the prevailing order imposed on them reflects the extent to which the structures of advanced capitalist societies have saturated every fibre of the being of those who live in them, including Marxists. To escape from these constraints, says Marcuse, we require

> a political practice of methodical disengagement from and refusal of the Establishment, aiming at a radical transvaluation of values. Such a practice involves a break with the familiar, the routine ways of seeing, hearing, feeling, understanding things so that the organism may become receptive to the potential forms of a non-aggressive, non-exploitative world.[84]

And, given the 'management of consciousness' (as he stunningly describes it) which occurs in the societies in question, Marcuse, like Feyerabend, emphasizes the importance of fantasy, of dreams, of the imagination and, more generally, of art as domains of knowledge with their own truth values.[85] These provide an alternative universe of perception and comprehension to that which is dominant.

An added dimension to these comments is provided by Thompson's discussion of William Morris' socialism. Morris' intention, says Thompson was to embody in the forms of fantasy alternative values sketched in an alternative way of life. And what distinguishes this enterprise is, exactly, its *open*, speculative, quality, and its *detachment* of the imagination from the demands of conceptual precision . . . In such an adventure two things happen: our habitual values (the 'commonsense' of bourgeois society) are thrown into disarray. And we enter into Utopia's proper and new found space: *the education of desire*. This is not the same as 'a moral education' towards a given end: it is, rather, to open a way to aspiration, to 'teach desire to desire, to desire better, to desire more, and above all to desire in a different way'.[86]

For these thinkers, in other words, the importance of breaking with the familiar is that one thereby becomes receptive to new possibilities, one opens out into a new domain. That the revolutionary project is *open*, and not guided by a clearly defined blueprint, has also been noted by Merleau-Ponty. In her account of his views, Kruks describes him as holding that

Free action does not have a clearly defined goal. A free action is 'open', it transcends the given, but it can be a leap into the unknown. Thus revolution is more likely to arise from a desire to change a restricting present than from a clearly defined conception of revolution, or of a post-revolutionary society . . . [Merleau-Ponty] compares the revolutionary movement to the work of an artist. Both are projects in which man asserts his freedom by transcending the present, but without knowing exactly where he is going. 'The revolutionary project, like the work of an artist, is an intention which itself creates its instruments and means of expression'.[87]

Compare this statement with a very similar claim made by Feyerabend which I have already quoted (see Ref. 66)

Creation of a *thing*, and creation plus full understanding of a *correct idea* of the thing, *are very often parts of one and the same indivisible process* and cannot be separated without bringing the process to a stop. The process is not guided by a well-defined programme, and cannot be guided by such a programme, for it contains the conditions for the realization of all programmes. It is guided rather by a vague urge, by a 'passion' (Kierkegaard).

Both Feyerabend and Merleau-Ponty stress the openness of the revolutionary project. But whereas Feyerabend says that it is guided by an urge or passion, Merleau-Ponty is more specific. It arises, he says, from a desire *to change the restricting present*. It involves, in Marcuse's words, 'methodical *disengagement* from and *refusal* of the Establishment'.

It is my contention that Feyerabend lands up advocating Dadaism because he fails to appreciate that Copernicus and Galileo did not simply *disengage* themselves from the Ptolemaic/Aristotelian system. They also *rejected* it, and *opposed* it. In other words, I accept that Copernicus and Galileo were committed to the possibility of a heliostatic cosmology, and to the view that the Earth was a planet. But I want to insist that no account of the tenacity with which they clung to those beliefs can afford to overlook the cardinally important fact that *they both believed that the alternative system of the universe was irredeemably flawed.*

This is quite explicit in Copernicus' Preface to *De Revolutionibus* in which he briefly surveys the prevailing state of his field and concludes that a radical change is called for. Summarizing his position Kuhn writes

> An honest appraisal of contemporary astronomy, says Copernicus, shows that the earth-centred approach to the problem of the planets is hopeless. The traditional techniques of Ptolemaic astronomy have not and will not solve that problem; instead they have produced a monster; there must, he concludes, be a fundamental error in the basic concepts of traditional planetary astronomy.[88]

Copernicus' commitment to his astronomical system was thus *also* a rejection of geocentric astronomy. Believing the latter to be beyond redemption he offered his alternative. Although it was in empirical difficulty as well, it did at least show that another approach to the problem of planetary motion was possible and it held out the hope that a more elegant theory than one relying on the cumbersome system of epicycles and deferrents may be forthcoming.

The case of Galileo is a little more difficult because Feyerabend quotes him as having said, in *The Assayer*, that both the Ptolemaic and the Copernican systems were 'surely false'. However, as Machamer has pointed out this comment was made by Galileo in the first book to be published under his name since the condemnation of Copernicanism in 1616.[89] It was, he suggests, an attempt by Galileo to appease the church authorities, and Feyerabend should not have taken what he said at face value. And indeed it is plain that Galileo was proceeding extremely cautiously in *The Assayer* because just a few pages earlier he accused an opponent of using foul play 'in order to get an opening through which he could somehow brand me a Copernican'.[90] In any case there is ample evidence elsewhere to suggest that Galileo believed the Aristotelian/Ptolemaic system to be mistaken. Note his combination of unhesitating rejection and tentative acceptance in the following remark, for example, made in 1615

> It is true that to prove that the appearances may be saved with the motion of the earth . . . is not the same as to prove this theory true in nature; but it is equally true, or even more so, that the commonly accepted system cannot give reasons for those appearances. *That system is undoubtedly false, just as . . . this one may be true.*[91]

There can be no doubt then that both Copernicus and Galileo were *opposed to* an Earth-centred cosmology, and that their conviction that it needed replacing was an essential component inspiring their defence of a heliostatic system. Copernicus believed that the problem of retrograde planetary motion *could not* be solved using Ptolemaic techniques. Galileo believed that the view that the Earth was stationary was *undoubtedly false*. On that account they rejected the orthodoxy of the day, and laid the foundations for a new conception of the universe which ultimately replaced it.

In the Introduction I pointed out that the 'liberal' conception of revolution is defective because it sees discontinuous transitions as being effected more or less suddenly and in one fell swoop. I suggested that an alternative conception is needed which detaches rejection from replacement. The revolutionary project, I remarked, is 'asymmetrical'. An established system is rejected, but it takes time to replace it with an alternative framework. Rejection and replacement, in other words, do not occur simultaneously. They are out of phase with each other.

What this means is that if the vision of the future is blurred at the moment of revolutionary rupture, the perception of the present is crystal clear. For Galileo the geostatic scheme was *undoubtedly* false. By contrast he claimed only that his preferred alternative *may* be true. It is this combination of conviction and tentativeness which *together* inform revolutionary commitment.

The revolutionary project is an open one. But openness to the future and the risks that that involves is combined with an unflinching rejection of the orthodoxy of the day. Feyerabend only focusses on the former aspect of the process. And since he rightly realizes that empirical success cannot provide revolutionary agents with the support they need to uphold their new views, he appeals to non-rational factors to explain their determination. I am suggesting that this is not, or at least need not be, necessary. There is an alternative source of revolutionary energy — the conviction that an established view is intrinsically defective, and incapable of accomodating further growth. It may ultimately be necessary to include a non-rational component in a theory of discontinuous transition. But it is premature to appeal to vague urges (à la Feyerabend) before developing a conception of rational change which integrates both dimensions of the revolutionary process.

Of course Feyerabend's failure to build the notion of theory rejection into his account of revolutionary commitment is consistent with the general drift of his argument and his espousal of Dadaism. He is not simply unhappy about rejecting *novel* theories on the grounds that they are empirically defective. In the wake of Lakatos' work he recognizes that, in general, it is unsound to use empirical inadequacy as a basis for rational theory rejection. Thus if one confines one's conception of rational theory choice to empirical considerations — in accordance with the principle of empiricism (Popper) — one has little option but to stress the importance of the non-rational, particularly at times of revolutionary change. From this it is but a short step to the view that all theory rejection and elimination must be coerced. 'Anything goes' then emerges as the only humane policy, its conservative implications notwithstanding.

I am fairly certain that in much of his work Feyerabend is simply drawing out the implications of the principle of empiricism without espousing it or its associated conception of rationality. Whether or not I am right it is, I think, clear that if theory rejection is to be incorporated into a conception of rational revolutionary commitment, that rejection has to be based on something more than empirical considerations.

An important clue to what this might be is provided in Kuhn's characterization of Copernicus' dissatisfaction with Ptolemaic astronomy. Copernicus, he says, came to believe that traditional astronomy not only *would not*, but actually *could not* solve the problem of retrograde planetary motion. In other words it was because he thought that it was *impossible* to deal adequately with retrograde motion within a geostatic framework that Copernicus developed his heliostatic alternative. The old is rejected because there is a growing awareness that it is impossible to work fruitfully within its confines any longer. The foundations of an alternative are laid, and an entire field of new possibilities comes into view. Revolutionary commitment, I want to suggest involves the fusion of these two components. It is because one believes that it is impossible to go on in the established way that one has the courage to reject it, and to move forward into a different future, exploring the possibilities inherent in an alternative form of life. In what follows I shall begin to explore the implications of these claims.

Notes

1. K.R. Popper, 'Replies to my critics' in P.A. Schilpp (ed), *The Philosophy of Karl Popper*, Vol. II (Open Court, La Salle, 1974), p.1069.
2. I. Lakatos and E. Zahar, 'Why did Copernicus' research program supersede Ptolemy's ?' in R.S. Westman (ed), *The Copernican Achievement* (University of California Press, Berkeley, 1975), p.367.
3. P.K. Feyerabend, 'How to be a good empiricist — a plea for tolerance in matters epistemological' in B. Baumrin (ed), *Philosophy of Science. The Delaware Seminar*, Vol. 2, 1962–3 (Interscience Publishers, New York, 1963), p.32.
4. Galileo Galileo, 'The Assayer', in S. Drake (trans), *Discoveries and Opinions of Galileo*, (Doubleday, New York, 1957), pp.237–8.
5. Galileo Galilei, *Dialogue Concerning the Two Chief World Systems — Ptolemaic and Copernican*, S. Drake (trans), (University of California Press, Berkeley, 1967), p.462.
6. See, for example, W.R.J. Shea, 'Galileo's claim to fame: the proof that the Earth moves from the evidence of the tides', *British Journal for the History of Science*, 5 (1970), 111–27.
7. S. Drake, in Galileo, *Dialogue, op.cit.* p.491.
8. S. Drake, *Galileo Studies*, (University of Michigan Press, Ann Arbor, 1970), p.72.
9. P.K. Feyerabend, *Against Method* (New Left Books, London, 1975), p.147.
10. P.K. Feyerabend, 'Problems of microphysics', in R.G. Colodny (ed), *Frontiers of Science and Philosophy* (Allen & Unwin, London, 1964), p.195.

11 A. Einstein, quoted in P.K. Feyerabend, *Against Method, op.cit.* p.18.
12 P.K. Feyerabend, in R.G. Colodny, *op.cit.* p.230.
13 N. Bohr, quoted in P.K. Feyerabend, 'Explanation, reduction, and empiricism', in H. Feigl and G. Maxwell (eds), *Minnesota Studies in the Philosophy of Science*, Vol. III (University of Minnesota Press, Minneapolis, 1962), p.43.
14 W. Heisenberg, quoted in P.K. Feyerabend, 'Problems of empiricism', in R.G. Colodny (ed), *Beyond the Edge of Certainty* (Prentice-Hall, New Jersey, 1965), pp.166–7.
15 E. Nagel, quoted in P.K. Feyerabend, in H. Feigl and G. Maxwell (eds), *op.cit.* p.33.
16 C.G. Hempel and P. Oppenheim, quoted in P.K. Feyerabend, in H. Feigl and G. Maxwell (eds), *op.cit.* p.34.
17 P.K. Feyerabend, 'On the improvement of the sciences and the arts, and the possible identity of the two' in R.S. Cohen and M.W. Wartofsky (eds), *Boston Studies in the Philosophy of Science*, Vol. III (Reidel, Dordrecht, Holland, 1967), p.405.
18 P.K. Feyerabend, 'Consolations for the specialist', in I. Lakatos and A. Musgrave (eds), *Criticism and the Growth of Knowledge* (Cambridge University Press, Cambridge, 1970), p.201.
19 P.K. Feyerabend, in H. Feigl and G. Maxwell (eds), *op.cit.* p.48.
20 For example, the observational consequences of classical and relativistic mechanics are indistinguishable from each other for bodies moving at velocities far smaller than the speed of light. As stated, this of course assumes that the two theories are not incommensurable with one another and that they share a common observation language. Incommensurability can readily be accommodated by distinguishing between (allegedly incommensurable) observation statements, and the measured values of parameters, or 'numbers'.
21 P.K. Feyerabend, in B. Baumrin (ed), *op.cit.* pp.22–3.
22 P.K. Feyerabend, 'Reply to criticism', in R.S. Cohen and M.W. Wartofsky (eds), *Boston Studies in the Philosophy of Science*, Vol.2 (Humanities Press, New York, 1965), p.223, (original emphasis suppressed).
23 P.K. Feyerabend, in B. Baumrin (ed), *op.cit.* p.6.
24 P.K. Feyerabend, in H. Feigl and G. Maxwell (eds), *op.cit.* p.50.
25 P.K. Feyerabend, *Against Method, op.cit.* p.46.
26 P.K. Feyerabend, 'Science: the myth and its role in society', *Inquiry*, 18 (1975), 176–7.
27 P.K. Feyerabend, 'Philosophy of science: a subject with a great past', in R.H. Steuwer (ed), *Minnesota Studies in the Philosophy of Science*, Vol. V (University of Minnesota Press, Minneapolis, 1970), p.173.
28 P.K. Feyerabend, in I. Lakatos and A. Musgrave (eds), *op.cit.* p.210.
29 P.K. Feyerabend, 'Logic, literacy and Professor Gellner', *British Journal for the Philosophy of Science*, 27 (1976), 387.
30 P.K. Feyerabend, 'Problems of empiricism, Part II', in R.G. Colodny (ed), *The Nature and Function of Scientific Theories* (The University of Pittsburgh Press, Pittsburgh, 1970), pp.277–8.
31 P.K. Feyerabend, 'From incompetent professionalism to professional-

ized incompetence — the rise of a new breed of intellectuals', *Philosophy of the Social Sciences*, 8 (1978), 41.
32 P.K. Feyerabend, in R.G. Colodny (ed), *Frontiers of Science and Philosophy, op.cit.* p.228.
33 L.O. Kattsoff, 'Ptolemy and scientific method', *Isis*, 38 (1947), 18, 19.
34 E.J. Dijksterhuis, *The Mechanization of the World Picture*, C. Dikshoorn (trans), (Oxford University Press, London, 1961), p.70.
35 Melanchthon, quoted in T.S. Kuhn, *The Copernican Revolution* (Harvard University Press, Cambridge, Mass., 1957), p.191.
36 Galileo, *Dialogue, op.cit.* p.328.
37 *Ibid.* p.256.
38 P.K. Feyerabend, *Against Method, op.cit.* pp.304–5.
39 P.K. Feyerabend, in H. Feigl and G. Maxwell (eds), *op.cit.* p.50.
40 P.K. Feyerabend, in R.S. Cohen and M.W. Wartofsky (eds), *op.cit.* p.405.
41 P.K. Feyerabend, in R.G. Colodny (ed), *The Nature and Function of Scientific Theories, op.cit.* p.227.
42 P.K. Feyerabend, *Against Method, op.cit.* p.155.
43 P.K. Feyerabend, 'Against method: outline of an anarchist theory of knowledge', in M. Radner and S. Winokur (eds), *Minnesota Studies in the Philosophy of Science*, Vol. IV (University of Minnesota Press, Minneapolis, 1970), p.129.
44 T.S. Kuhn, *op.cit.* p.171.
45 P.K. Feyerabend, *Against Method, op.cit.* p.26.
46 H.V. White, 'Foucault decoded: notes from underground', *History and Theory*, 12 (1973), 27. On the question of transition, however, see M. Foucault, *The Order of Things* (Tavistock, London, 1974), p.xiii.
47 P.K. Feyerabend, *Against Method, op.cit.* p.88.
48 See, for example, P.K. Feyerabend, *Against Method, op.cit.* pp.55, 102.
49 *Ibid.* p.73.
50 An example of a natural interpretation which is being challenged by contemporary feminists is the claim that temperamental differences between the sexes are biologically based. The difficulty for them, as A. Oakley writes in *Sex, Gender and Society* (Temple-Smith, London, 1972), p.49, is that 'Only the *everyday observation* [my emphasis] of men and women in society is needed to 'prove' that differences of personality follow the biological differences of sex. Men are more aggressive and independent than women; they are braver, more outgoing and extroverted, confident in their own ability to control and manipulate the external environment. Women are more sensitive and perceptive in their relationships with other people; they are more dependent on these relationships. They are introverted and domesticated and emotionally labile'. Feminists argue that this is just one of many attempts to retain established sex roles by claiming that they are biologically based. The conservative argument is persuasive because 'in Western societies today sex *is* an organising principle of social structure, and . . . it plays a great part in determining social roles' (A. Oakley, *op.cit.* p.150). The 'facts' are accordingly heavily biassed in favour of biological determinism. In this connection see also J. Krige, 'What's so

great about facts ?', in J. Irvine, I. Miles and J. Evans (eds), *Demystifying Social Statistics* (Pluto Press, London, 1979), pp.53–62.
51 Galileo, *Dialogue, op.cit.* p.171.
52 *Ibid.* p.171.
53 P.K. Feyerabend, *Against Method, op.cit.* p.89.
54 P.K. Feyerabend, *Against Method, op.cit.* p.89. Another reason why Feyerabend thinks that the non-operative character of shared motion was an inadequate theory is that he thinks that Galileo's kinematic relativism is not consistent (*Against Method, op.cit.* p.85). This very prevalent view has been challenged by Brown, in his 'Galileo, the elements and the tides', *Studies in the History and Philosophy of Science*, 7 (1976), 337–51. Feyerabend's treatment of the tower argument has been criticized by McMullin in R.H. Steuwer, (ed), *Minnesota Studies*, Vol. 5, pp.34–41. Feyerabend has replied (*Against Method, op.cit.* p.85).
55 T.S. Kuhn, *op.cit.* p.219.
56 Quoted in P.K. Feyerabend, *Against Method, op.cit.* p.123.
57 P.K. Feyerabend, *Against Method, op.cit.* p.142. Machamer has criticized this argument in a lengthy article (P.K. Machamer, 'Feyerabend and Galileo: the interaction of theories and the reinterpretation of experience', *Studies in the History and Philosophy of Science*, 4 (1973), 1–46. He relies in part on an interpretation of Galileo's lunar drawings made by Righini. Gingerich has rejected Righini's interpretation (O. Gingerich, 'Dissertatio cum Professore Righini et Sidereo Nuncio' in M.L. Righini Bonelli and W.R. Shea (eds), *Reason, Experiment, and Mysticism in the Scientific Revolution* (Science History Publications, New York, 1975), pp.77–88. Feyerabend has replied to Machamer in *Against Method, op.cit.* Appendix 2.
58 P.K. Feyerabend, *Against Method, op.cit.* p.143.
59 *Ibid.* p.167.
60 Galileo *Dialogue, op.cit.* p.416.
61 P.K. Feyerabend, *Against Method, op.cit.* p.154.
62 *Ibid.* p.32.
63 *Ibid.* p.189.
64 P.K. Feyerabend, in M. Radner and S. Winokur (eds), *op.cit.* p.103.
65 P.K. Feyerabend, 'Incompetent professionalism', *op.cit.* pp.43–4.
66 P.K. Feyerabend, *Against Method*, op.cit. p.26.
67 H.V. White, *op.cit.* p.39.
68 P.K. Feyerabend, in R. Steuwer (ed), *op.cit.* p.173.
69 P.K. Feyerabend, *Against Method, op.cit.* p.33.
70 *Ibid.* p.189.
71 P.K. Feyerabend, in M. Radner and S. Winokur (eds), *op.cit.* p.107.
72 P.K. Feyerabend, *Against Method, op.cit.* p.191.
73 P.K. Feyerabend, 'Outline of a pluralistic theory of knowledge and action', in S. Anderson, (ed), *Planning for Diversity and Choice* (MIT Press, Cambridge, Mass., 1968), p.276.
74 R. Williams, 'Base and superstructure in Marxist cultural theory', *New Left Review*, No. 82 (1973), 11.
75 P.K. Feyerabend, in I. Lakatos and A. Musgrave (eds), *Criticism and the Growth of Knowledge, op.cit.* p.210.

76 P.K. Feyerabend, in R.G. Colodny (ed), *The Nature and Function of Scientific Theories, op.cit.* p.227.
77 P.K. Feyerabend, *Against Method, op.cit.* pp.153, 154.
78 P.K. Feyerabend, 'The theatre as an instrument of the criticism of ideologies', *Inquiry*, 10 (1967), 302.
79 *Ibid.* p.306.
80 P.K. Feyerabend, *Against Method, op.cit.* p.154.
81 P.K. Feyerabend, in M. Radner and S. Winokur (eds), *op.cit.* p.110.
82 P.K. Feyerabend, *Against Method, op.cit.* p.300.
83 H. Marcuse, 'Herbert Marcuse on the need for an open Marxist mind', *The Listener*, 9 February 1978, p.169.
84 H. Marcuse, *An Essay on Liberation* (Penguin, Harmondsworth, 1972), pp.15–6.
85 H. Marcuse, *Eros and Civilization*, (Sphere Books, London, 1969), Chapter 7.
86 E. Thompson, 'Romanticism, Utopianism and Moralism: the case of William Morris', *New Left Review*, No.99 (1976), 97.
87 S. Kruks, 'The philosophy of Merleau-Ponty', *Radical Philosophy*, No.11 (Summer, 1975), p.23.
88 T.S. Kuhn, *op.cit.* p.138.
89 P.K. Machamer, 'Feyerabend and Galileo: the interaction of theories, and the reinterpretation of experience', *Studies in the History and Philosophy of Science*, 4 (1973), 3.
90 Galileo Galilei, *The Assayer*, in S. Drake and C.D. O'Malley (trans), *The Controversy on the Comets of 1618* (University of Pennsylvania Press, Philadelphia, 1960), p.180.
91 Galileo Galilei, quoted in S. Drake (trans), *op.cit.* p.169, my emphasis.

5 STRUCTURES AND IMPOSSIBILITIES

5.1 Falsehood and Impossibility

LOOKING back over the work of the three philosophers whom I have discussed in some depth, one point emerges with particular clarity. It is that there is a steady drift, from Popper through Lakatos to Feyerabend, away from the idea that one can *reasonably reject* a system of thought, replacing it with a more adequate view. At one extreme, for Popper, scientific knowledge grows by the method of conjecture, refutation, and theory elimination. At the other extreme, for Feyerabend, anything goes, and growth involves the parallel development of a plurality of world views each embodying its own canons of rationality. As far as he is concerned, the attempt to formulate rational, 'objective' criteria for choosing between alternative world views is doomed to failure. When it comes to making such decisions all that we can rely on are 'aesthetic judgements, judgements of taste, metaphysical prejudices, religious desires, in short, *what remains are our subjective wishes*'.[1]

In the previous chapter I showed how Feyerabend's analysis of Galileo's behaviour was informed by this view. He argues that Galileo could not reasonably have rejected an Earth-centred cosmology because he thought that it was false. Rather, Galileo's preference for Copernicanism was ultimately based on subjective grounds, which he subsequently rationalized. It is considerations of this kind which have led Feyerabend to defend Dadaism.

The idea that the initial choice between the geostatic and the heliostatic conceptions of the universe was not made 'objectively' is also prevalent in several influential treatments by historians of the Copernican Revolution. Typically, de Solla Price remarks that, concerning the choice between Copernicus' and Ptolemy's mathematical theories of planetary motion, 'there was no possible observation that could decide whether one was true and the other false'.[2] Kuhn explains why. The two theories are observationally equivalent, he says, because 'the motion of the sun has simply been transferred to the earth'.[3]

That granted, on what basis was a choice to be made between the theories ? Copernicus' alternative was neither more economical nor more precise than its predecessor. Following Copernicus' lead, both Price and Kuhn stress that whereas in Ptolemaic astronomy planets were treated separately and individually, Copernicus invented a mathematical planetary system. The adherents of traditional astronomy, said Copernicus, had not been able 'to discern or deduce the principal thing — namely the shape of the Universe and the unchangeable symmetry of its parts'.[4] In his heliostatic system, by contrast 'the orders and magnitudes of all stars and spheres, nay the heavens themselves, become so bound together that nothing in any part thereof could be moved from its place without producing confusion of all the other parts and of the Universe as a whole'.[5] It was this coherence and geometric harmony which really counted in favour of Copernicanism. Copernicus recognized this, recognized, says Kuhn, that 'the real appeal of sun-centred astronomy was aesthetic rather than pragmatic. To astronomers the initial choice between Copernicus' system and Ptolemy's could only be a matter of taste'.[6]

Where Feyerabend speaks of subjective wishes, then, Kuhn speaks of matters of taste. Both of them draw a rigid distinction between aesthetic considerations, dealing with coherence and system, and questions of truth and falsehood. Superimposed on this questionable dichotomy,[7] the historians draw a further distinction between the mathematical aspects of Copernicus' theory and its physical dimension, concerned with the Earth's motion.[8] Thus Price regards Copernicus' assumption that the Earth is a planet to be a 'purely philosophical device'. He (Price) is not concerned 'with the philosophical battles over the systems but with the purely scientific ones (in the modern sense) about which theories corresponded better with the observations'.[9] Similarly Kuhn divides the Copernican Revolution into two distinct phases which allegedly followed sequentially on one another. After an initial non-rational commitment to the heliostatic system (informed by aesthetic considerations), mathematical astronomers set out to resolve the ensuing dynamical problems related to the earth's alleged movement. In Kuhn's words

> Those whom Copernicus converted to the concept of a moving earth began their research from the point at which Copernicus had stopped. Their starting point was the earth's motion, which was all they necessarily took from Copernicus, and the problems to which they devoted themselves were not the problems of the old astronomy, which had occupied Copernicus, but the problems of the new sun-centred astronomy, which they discovered in the *De Revolutionibus*.[10]

This sounds very much as though their arguments for a heliostatic system were simply rationalizations for a prior commitment to a theory which its proponents found pleasing. Kuhn, like Feyerabend, is well on the way to turning Copernicans into Dadaists.

We can now see how it is that the idea that the Ptolemaic system was rejected because it was false is theorized out of historical accounts of the Scientific Revolution. Firstly, a distinction is drawn between the physical and the mathematical (i.e., optical) aspects of the Earth's motion. Whether or not the Earth really moved — this is treated as if it were a cosmological question, an issue that was settled by Aristotelian philosophers, and which was *ipso facto* not 'scientific'. Focussing then on the Ptolemaic and Copernican theories as alternative mathematical systems advanced to save the celestial appearances, it is argued that observation could not be used to choose unambiguously between them (at least until telescopic evidence was forthcoming). Drawing a distinction between system or coherence, on the one hand, and truth and falsehood on the other, it is then concluded that the choice between the two theories must have been arbitrary, a matter of taste, and that the transition from one to the other was a 'conversion experience' (Kuhn).

There is some justification in Copernicus' work for adopting these distinctions. Firstly, as we have seen, in the Preface to *De Revolutionibus* Copernicus did emphasize the superior coherence of his theory, apparently valuing aesthetic considerations above, and as distinct from, empirical adequacy. Furthermore, and related to this, fearing that he would be 'hissed off the stage' for asserting that the Earth moves, Copernicus turned to mathematicians for support. He thus rather one-sidedly suggested in the Preface that it was solely as a mathematical hypothesis that his theory should be assessed. To that extent he was simply falling back on the age-old schism between the domains of the astronomer and of the physicist which had dominated Greek thought. The task of the former was primarily that of saving the celestial phenomena using the appropriate mathematical and arithmetical devices. Physicists, or natural philosophers, on the other hand, aimed to establish the true constitution of the universe, its nature and its essence.[11]

Now it may be true that Copernicus relied upon certain dubious distinctions to protect himself from censure. But it is just as certain that Galileo was not prepared to hide behind smokescreens of this kind. As a confrontation with the Church became imminent, Cardinal Bellarmine spelt out the official position

> For to say that assuming the earth moves and the sun stands still saves all the appearances better than eccentrics and epicycles is to speak well. This has no danger in it, and it suffices for mathematicians. But to wish to affirm that the sun is really fixed in the center of the heavens . . . and that the earth . . . revolves very swiftly around the sun, is a very dangerous thing, not only by irritating all the theologians and scholastic philosophers, but also by injuring our holy faith and making the sacred Scripture false.[12]

If he had been alive, Copernicus may have been prepared to cooperate with Bellarmine, at least in public. Galileo was not. Mathematical astronomers, he said, may be content to save the appearances with devices like eccentrics and epicylces. Such devices, however

> are not retained by philosophical astronomers who, going beyond the demand that they somehow save the appearances, seek to investigate the true constitution of the universe — the most important and most admirable problem that there is. For such a constitution exists; it is unique, true, real and could not possibly be otherwise; and the greatness and nobility of this problem entitle it to be placed foremost among all questions capable of theoretical solution.[13]

And so, in a point-by-point reply to Bellarmine's letter, preserved in his notes, Galileo made it clear that he was not to be intimidated

> It is true [he wrote] that to prove that the appearances may be saved with the

motion of the earth . . . is not the same as to prove this theory true in nature; but it is equally true, or even more so, that the commonly accepted system cannot give reasons for those appearances. That system is undoubtedly false, just as . . . this one may be true.[14]

One thing was certain: between the Two Chief World Systems, Ptolemaic and Copernican, a decision must be taken which, wrote Galileo

ultimately should not be ambiguous, since one of the arrangements must be true and the other false . . . . For between motion and rest, which are contradictories, there is no middle ground (as if one might say the earth neither moves nor stands still; the sun and the stars do not move and do not stand still).[15]

Whatever Copernicus' aims were, then, Galileo saw himself as struggling to establish the true constitution of the universe. What's more, he rejected as false Ptolemaic astronomy or, more precisely, the Ptolemaic/Aristotelian system of the world. Any adequate account of his achievement must take these features of it into consideration.

Galileo repeatedly insisted that there could be only one true account of reality, and that evidence which appeared to support a false view must necessarily speak in favour of the true conception as well. Typically, comparing the evidence for and against the Ptolemaic and Copernican systems, he remarked that

if one of these positions is true and the other necessarily false, it is impossible for any reason, experiment, or correct argument to be found to favour the false one, as none of these things can be repugnant to the true position. Therefore a great disparity must exist between the reasons and arguments that are adduced by the one side and by the other for and against these two opinions.[16]

This was simply because there was only one possible way in which the universe could be constituted (see Ref. 13) and a true account of it reflected that constitution. Correlatively, a false account represented an impossible reality, a 'reality' that could not possibly be.

According to Galileo, the Aristotelians regarded the Copernican hypothesis in this light. His mouthpiece, Salviati, opens the debate in the *Dialogue Concerning the Two Chief World Systems* in the following vein

SALV. Yesterday we resolved to meet today and discuss as clearly and in as much detail as possible the character and the efficacy of those laws of nature which up to the present have been put forth by the partisans of the Aristotelian and Ptolemaic positions on the one hand, and by the followers of the Copernican system on the other. Since Copernicus places the earth among the movable heavenly bodies, making it a globe like a planet, we may well begin our discussion by examining the Peripatetic steps in arguing the impossibility of that hypothesis; what they are, and how great is their force and effect.[17]

Salviati, in other words, unambiguously asserts that for a Peripatetic it is not simply false that the Earth moves; it is actually *impossible* for it to do so.[18] My aim in this chapter is to elucidate the implications of this opening gambit. By 'rationally reconstructing' Aristotelian thought, spelling out the associated conception of reality, I shall explain why it was that the Aristotelians believed that it was impossible for the Earth to move circularly. Conversely, I shall show why it was that Galileo believed that Copernicanism could not simply be grafted onto the edifice of Peripatetic thought. The defense and the articulation of the heliostatic system involved the rejection and, ultimately, the elimination of Aristotelian physics. The processes whereby it was replaced with an entirely new system of thought and a new conception of nature will be discussed at some length in Chapter 6.

5.2 The Structure of Aristotelian Thought

At the very start of the *Dialogue* Galileo makes it plain that one of his aims is to isolate and to challenge the fundamental tenets of Aristotelianism. This is clear from Salviati's opening statement of which I have quoted only a fragment so far. Continuing where the previous quotation left off, Salviati explains how one is to go about 'examining the Peripatetic steps in arguing the impossibility' of the Earth's movement

SALV. For this it is necessary to introduce into nature two substances which differ essentially. These are the celestial and the elemental, the former being invariant and eternal; the latter, temporary and destructible. This argument Aristotle treats in his book *De Caelo*, introducing it with some discourses dependent upon certain general assumptions, and afterwards confirming it by experiments and specific demonstrations.[19]

Salviati, then, attributes the impossibility of conceding that the Earth moves (circularly) to the restrictions imposed by a number of Aristotelian assumptions which, in conjunction with sense experience, generated a rigid distinction between the terrestrial and the celestial. One of Galileo's main aims in the *Dialogue* was to identify these basic principles. By criticizing them he simultaneously created the space in which he laid the foundations for a new conceptual framework compatible with Copernicanism.

The first 'general assumption' which Salviati immediately introduces is Aristotle's belief in the 'completeness and perfection of the world'.[20] This conception of the universe as a cosmos or ordered whole was a central feature of Peripatetic thought. And, as Koyré notes

Whole, cosmic order; these ideas imply that everything in the Universe is, or ought to be, distributed and arranged in a highly determinate manner; that things are not indifferent as to whether they be here or there, but that, on the contrary, everything in the Universe has its proper place in conformity with its nature. A place for each thing and everything in its place; the idea of 'natural place' is the theoretical expression of this necessity in Aristotelian physics.[21]

For the Aristotelians the idea that things had natural places in which they would reside unless and until they were forcibly disturbed rendered motion and change both necessary and intelligible. In fact, one of Aristotle's outstanding achievements was his attempt to construct a coherent science of change. To this end he articulated a set of concepts which integrated his physics tightly into his cosmology.

Against Parmenides who had held that being was eternal and immutable, Aristotle argued that far from being incompatible with being, the origins of change lay in being itself.[22] For, he said, any concrete substance or subject has both an *actual* existence, by virtue of which it is here and now, and a *potential* existence, constituted by those of its unrealized possibilities which it is capable of actualizing. The totality of its actual and potential qualities is embodied in its form, which is unfulfilled precisely to the extent that some possibilities are not actualized. In other words, far from resting content in a state of relative non-being every substance or subject experiences that condition as one of *privation*. By defining potentiality as privation, as Clavelin puts it, 'Aristotle was simply asserting that every concrete being is constantly impelled, from the inside as it were, to make all sorts of changes, and so acquires a sense of expectation'.[23] It is because potential existence is experienced as a privation that movement and change occur in the first place, having the indispensable function of actualizing unrealized possibilities.

In the light of later developments in classical mechanics, to be discussed in more detail in Chapter 6, the most striking aspect of Aristotle's conception of motion was his treatment of it as a *transitory process*. This was a consequence of his regarding it as the means whereby the potential is transformed into the actual; on achieving the latter, motion lost its *raison d'être*. For Aristotle motion was inconceivable without a *terminus a quo*, from which starting point it propelled a subject towards a *terminus ad quem*. Once this end point was reached, motion and change ceased. In other words, motion was not an independent state of a body, existing in its own right; it had a carefully specified ontological function — the actualization of the potential — and apart from that it was unintelligible.

The conceptual apparatus outlined above was invoked by Aristotle to account for motion and change in general. Local motion, or change of place, was simply one particular kind of change which he included within a more comprehensive categorization, but to which he attributed especial importance. In the case of local motion the *a priori* need for precisely specified termini led to the demand for the existence of predetermined places between which movement occurred. It is here that Aristotle's cosmology intersects with his science of change. By postulating that 'everything in the Universe has its proper place in conformity with its nature' (see Koyré, above) it provided local motion with those (natural) termini without which it would have been meaningless.

According to Aristotelian cosmology, the universe is finite and spherical and is characterized by six directions (i.e., up and down, right and left, and forward and backward). Of these the first pair was the most important for his discussion of local motion. Aristotle insisted that the directions 'up' and 'down' could be 'objectively' defined for they also depended on the way in which bodies moved. More specifically they indicated the directions of two of the three motions which Aristotle took to be 'simple'. Salviati explains the connection. Informed by 'ideal', 'architechtonic' considerations, he says, Aristotle calls circular and rectilinear motion simple 'because of all lines only the circular and the straight are simple'. Then he

> defines among the simple motions one, the circular, to be that which is made around the center; and the other, the straight, to be upward and downward — upward, that which goes from the center; and downward whatever goes toward the center. And from this he infers it to be necessary and proper that all simple motions are confined to these three kinds; namely, toward the center, away from the center, and around the center. This answers, he says, with certain beautiful harmony to what has been said previously about the body; it is perfect in three things and its motion is likewise.[24]

To be intelligible, local motion required termini; Aristotelian cosmology provided them, and restricted the kinds of motion compatible with the perfection of the universe to three. These three simple motions are responsible for the order of the world. They preserve the organization and harmony of the cosmos, and that by distributing bodies in their natural places, and by restoring them to those places if and when such bodies have been violently disturbed.

If bodies are to move naturally or spontaneously to their termini to preserve order, it is necessary to endow them with inherent tendencies which will drive them upward or downward as the case may be. The qualities of lightness and heaviness serve this purpose in Aristotle's scheme of thought. Light bodies, he said, naturally tend to move towards the circumference of the world, and heavy bodies to move downwards towards its centre. In fact, Aristotle asserted, 'this is precisely what it means for them to be light or heavy — namely that they tend upward or downward, respectively'.[25] In other words Aristotle regarded lightness and weight to be absolute qualities of bodies, and went as far as to criticize his precursors for treating them as relative terms and for thus failing to base the 'up' and the 'down' on natural foundations.

I mentioned above that natural motions served to restore to their natural places bodies which had been violently removed from them. This distinction between natural and unnatural or violent motions can now be clarified: violent motions are those which compel bodies to move in ways which are at variance with their inherent tendencies. For example, to propel a heavy body upwards (say, to throw a stone) is to force it to carry out a movement which violates its intrinsic tendency to move downwards towards the centre of the universe. The resistance which the body has to being (violently) disturbed, and the spontaneous tendency which it shows to return (naturally) to a state of rest are both a consequence of its participation in the order of the cosmos. That order is an essentially static one: in moving to its natural place a body actualizes its form and so, once there, it will tend to persist in that state. Thus whereas for Aristotle motion was a transitory process, rest was the 'natural' state in which a body persisted or to which it spontaneously returned. As Clavelin notes, the priority which rest enjoyed over motion in this conceptual scheme 'reveals the extent to which Aristotle's theory of local motion depends on his cosmological premises, and also the extent to which the construction of classical mechanics was dependent on the demolition of that cosmology'.[26]

To the directions 'up, and 'down' were linked the 'light' and the 'heavy'. Superimposed on this base and informed by Aristotle's theory of simple motions was his doctrine of the elements. There are, said Aristotle, as many primary elements as there are simple motions. Rectilinear motions towards and away from the centre of the universe are the simple motions that we encounter in the terrestrial zone. Corresponding to each is an element, one of which (Earth) has the inherent tendency to move downwards and the other of which (Fire) moves naturally upwards. For completeness, and to save the appearances, Aristotle added to these two further elements of a mixed nature which also moved rectilinearly — Water and Air.

The natural places of these four elements comprised concentric rings around the centre of the world, which were arranged outwards from Earth through Water and Air to Fire. Fire came to rest in an orb beneath the surface of the Moon. Through the generative and corruptive influence of the Sun these layers were continually disturbed; by moving rectilinearly back towards their natural places the several elements continuously struggled to restore cosmological harmony.

It remained for Aristotle to identify that element that moved naturally in a circle, the other (geometrically) simple trajectory. Now rectilinear motion is associated with bodies that move towards and away from the centre of the world and which are heavy and light, respectively. Since in circular movement bodies neither recede from nor advance towards the centre, the element that executes this motion can be neither heavy nor light. Furthermore, in rectilinear motion bodies move towards clearly specified termini in which they actualize their form. However, any point on a circular path is equivalent to any other; accordingly motion along a circle cannot be appropriate to a body that is striving towards a goal, nor therefore to a body that experiences privation. In other words bodies that move circularly already exist in the fulness of their being, they are pure actuality. And since only celestial bodies are subject to sustained circular motion it follows that they comprise a substance, which — as Galileo puts it — is believed to be 'ingenerable, incorruptible, inalterable, invariant, and in a word free from all mutations except those of situation, and accordingly to be a quintessence most different from our generable, corruptible, alterable bodies'.[27]

The importance which Galileo attached to the split between the terrestrial and the celestial cannot be overestimated. As we have seen it is the point at which he breaks into the structure of Aristotelian thought in the very opening lines of the *Dialogue*. To understand why the Peripatetics hold that it is impossible for the Earth to move circularly, he says, 'it is necessary to introduce into nature two substances which differ essentially. These are the celestial and the elemental, the former being invariant and eternal; the latter being temporary and destructible' (See Ref. 19, Section 5.2). This distinction, in turn, is generated by more deepseated demands of Aristotelian cosmology. It is the dichotomy between rectilinear and circular motion, says Galileo, which

is the cornerstone, basis, and foundation of the entire structure of the Aristotelian universe, upon which are superimposed all other celestial properties — freedom from gravity and levity, ingenerability, incorruptibility, exemption from all mutations except local ones, etc. All these properties he attributes to a simple body with circular motion. The contrary qualities of gravity or levity, corruptibility, etc., he assigns to bodies naturally movable in a straight line.[28]

As Salviati recognized, for the Peripatetics, 'the earth cannot move circularly because of its corruptibility and generability'.[29]

In this section I have explored the complex web of interlocking doctrines and arguments which underpin this assertion. To summarize them there is little point in trying to improve on Clavelin's excellent synthesis, through which the majesty of Aristotelian thought shines in all its splendour

> Combining sense experience with the assumption that only the circle and the straight line represent perfect geometrical lines, that is, lines whose successive parts, when superposed, can be made to fit precisely, Aristotle was led to posit that simple local motion can appear in two forms only: the circular and the rectilinear. Then, harmonising this assumption with the *a priori* orderly structure of the cosmos, he was able to infer the existence of three simple natural motions: straight upward motion away from the center, straight downward motion toward the center, and circular motion about the center. From the further assumption that each of these motions is appropriate to a different element, Aristotle concluded that the number of elements must be limited. Thus while earth and fire (to which he later added water and air) provided the physical foundations and accounted for the number of rectilinear motions, a fifth element, the ether, was the basis of simple circular motion. Moreover, simple natural motions held the key not only to the possible number of elements but also to some of their most essential properties. Thus, because all its successive parts are alike, natural circular motion is perfect and hence confined to bodies in actual existence, and as such ingenerable and incorruptible; while rectilinear motion, being indeterminate in itself and definable only be reference to external factors, is typical of imperfect bodies, that is, bodies possessed of gravity, levity, generability, and corruptibility. This explains why terrestrial bodies, which alone are generable, corruptible, and changeable move in straight lines, while celestial bodies, which are ingenerable, incorruptible, and constant, move in circles.[31]

All that needs to be added to this is the Aristotelian view that 'it is natural for the whole to be in the place toward which the part has a natural motion'.[31] The earthy moves naturally in a straight line towards the centre of the universe. There the whole, the Earth, will remain at rest, immobile in its natural place, until violently disturbed. It was impossible for the Earth to move perpetually describing circular motions. Copernicanism must, therefore, be false.

5.3 Structure and Natural Necessity

The coherence of Aristotelian thought can easily lead one to suppose that it was an intricate conceptual web spun solely by speculative thought. But this is not so. It was a system that was consciously intended to accommodate observable features of the world. More specifically, its very intimate links with everyday perceptual experiences have been stressed by Koyré and by Kuhn. Koyré has noted how 'natural' it seems for heavy bodies to fall to Earth and for fire to rise; if they failed to do so we would be puzzled, and would seek an explanation for their 'unnatural' behaviour. Aristotelian physics, he says, is a theory which takes its starting point in the data of common sense and submits them to 'a systematic, extremely coherent and rigorous elaboration'.[32] Kuhn has made a similar point. Part of the authority of the Aristotelian corpus, he says, lay in the brilliance of its original ideas, and part was derived from its vast range and logical coherence. But the primary source of its tenacity, he suggests, lay in Aristotle's ability 'to express in an abstract and consistent manner many spontaneous perceptions of the universe which had existed for centuries before he gave them a logical verbal rationale'.[33]

Koyré and Kuhn have undoubtedly isolated an important aspect of Aristotle's thought. For example, in the now familiar opening statement of the *Dialogue* Salviati emphasizes that the split between the terrestrial and the celestial is both derived from certain fundamental principles and based on observational evidence. And when Galileo a little later suggests that the natural motion of earthy matter in an ordered universe is not rectilinear at all, but circular, Simplicio — the Peripatetic spokesman — is incredulous and says

SIMP. Now who is there so blind as not to see that earthy and watery parts, as heavy things, move naturally downward — that is to say toward the center of the universe, assigned by nature itself as the end and terminus of straight motion *deorsum* ? Who does not likewise see fire and air move directly upward toward the arc of the moon's orbit, as the natural end of motion *sursum* ?[34]

Although it is quite true that Aristotelian cosmology 'reflected' certain gross features of everyday, commonsensical observation, there is also a far more complex relationship between it and the observable phenomena. This emerges clearly in Galileo's discussion of the tower argument for the immobility of the Earth.

Galileo regarded the tower argument, which I discussed briefly in Chapter 4, to be of a particularly persuasive type drawn from 'earthy experiments'. This argument, it may be remembered, relies on the observed fact that in free fall heavy bodies drop by a straight and vertical line (to a good approximation) to the surface of the Earth. This, so it was said, proves that the Earth does not rotate diurnally on its own axis. For if it did, a stone dropped from a tower would not fall alongside it. It would describe an arc through space and it would strike the ground some hundreds of yards to the west of the tower's base. Galileo is not satisfied. Through Salviati he asks Simplicio what kind of trajectory through space the body would have to follow to save the appearances on a moving Earth. The Aristotelian has no difficulty in replying: it would be a compound of two motions, 'one with which it went from top to bottom, and another one needed for following the path of the tower'.[35] But if that be granted, then it does not follow that the Earth is stationary simply because a stone released from the top of a tower is seen to graze its side. Another assumption is required, and Simplicio spells it out as follows

SIMP.   For to expect the rock to go grazing the tower if that were carried along by the earth would be requiring the rock to have two natural motions; that is, a straight one toward the center, and a circular one about the center, which is impossible.

SALV.   So Aristotle's defence consists in its being impossible, or at least in his having considered it impossible, that the rock might move with a motion mixed of straight and circular. For if he had not held it to be impossible that the stone might move both toward and around the center at the same time, he would have understood how it could happen that the falling rock might go grazing the tower whether that was moving or standing still, and consequently he would have been able to perceive that this grazing would imply nothing as to the motion or rest of the earth.[36]

Simplicio, then, has taken it for granted that earthy material cannot execute mixed straight-and-circular motions. In fact, he thought that it was impossible for it to do so. As far as he is concerned, when heavy bodies fall freely from a point above the surface of the Earth, they *must* descend along a straight line to the centre of the universe: the observed connection between the act of releasing a stone from the top of a tower and its falling directly to the base is not a contingent one. This is because, for the Aristotelians, a heavy terrestrial body like a rock is just the kind of thing which moves naturally downwards, restoring order and harmony to the universe by returning to the natural place in which its essence is actualized.

In Section 5.2 I spelt out the full conceptual structure in which assertions like that of Simplicio's which I have just quoted are embedded. However, it is now clear that the impossibility of which he (and Galileo) spoke, and the necessity which is its obverse is not simply a conceptual or logical impossibility (or necessity). Rather, it is what the contemporary proponents of 'scientific realism'[37] call natural necessity. Realists reject Hume's anthropocentric view that the idea that necessary connections hold between successive, observable events derives from the intensity of the associated impressions in the mind of the observer. They insist that, on the contrary, there are necessary connections between matters of fact which subsist independently of their being observed.

Such natural necessities, as they are called, hold between things, events, states of affairs, and so on, by virtue of their natures. They are reflected in the logical connections between the concepts, which are knotted together to form the structures with which we try to grasp reality. These conceptual structures embody a particular conception of nature asserting, for example, that the earthy is that which falls naturally 'by a straight and vertical line' to the centre of the universe.

If the earthy is characterized in this way it follows that there is not simply a contingent connection between the releasing of a stone from the top of a tower and its falling perpendicularly. Rather, the stone *must* follow this trajectory. The necessity in question is natural necessity.

These regular relations refer beyond themselves to the existence of deepseated structural mechanisms of which they are the surface appearances. 'The world', writes Bhaskar, 'consists of mechanisms not events. Such mechanisms combine to generate the flux of phenomena that constitute the actual states and happenings of the world'.[38] And as Keat and Urry point out, 'in describing these mechanisms and structures we will often, in effect, be characterizing the "nature", "essence", or "inner constitution" of various types of entity'.[39] Typically, in our case it is in the nature of heavy bodies that they should return by the shortest path to their 'natural' places.

We are now in a better position to appreciate the full import of Salviati's remark at the start of the *Dialogue* to the effect that, for the Peripatetics, it is impossible that the Earth should be placed among the movable heavenly bodies, and that it be treated as a planet. As he put it later when rephrasing one of the arguments for the Earth's immobility put forward by Simplicio

> SALV. If I remember correctly, Simplicio's first argument was this: The earth cannot move circularly, because such a motion would be a forced one and therefore not perpetual. The reason that it would be forced was that if it were natural, the earth's parts would also naturally move in rotation, which is impossible because the nature of these parts is to be moved downward in a straight line.[40]

For the Aristotelians, in other words, it was not simply false that the Earth rotated diurnally on its own axis and annually about the Sun; nor was the statement that it did so meaningless. Rather, they believed that it *could not* behave in that way, and that the assertion that it did *could not* be true. These possibilities, they held, were excluded by the nature of the object of investigation itself. The universe was structured in such a way as to make it impossible for the Earth to be anything other than stationary at its centre: this was a cosmological imperative. Aristotelian thought reflected the structure of the object which it represented. As such it was the only possible account of the world which could be true.

Galileo well appreciated that the assertion that it was impossible for the Earth to move circularly was enmeshed in a tightly knit conceptual scheme, each element of which derived its full significance through its relationship with the others and with reality. Accordingly he realized that, to exploit the full implications of Copernicanism, he would have to demolish the entire edifice of Aristotelian thought and its associated conception of nature. It was necessary for him to reconstitute the object of scientific enquiry afresh, and to generate an appropriate conceptual scheme which was cognitively adequate to it. The emergence of modern science was not the result of an act of poetic intuition (Popper), or of a gestalt-switch (Kuhn), or of an arbitrary and inspired guess expressive of personal inclination (Feyerabend). Rather, as Koyré has stressed, it was the outcome, at least in part, of a tremendous intellectual effort by

> some of the deepest and mightiest minds every produced by mankind . . . [who] had, to begin with, to reshape and to re–form our intellect itself; to give to it a series of new concepts, to evolve a new approach to being, a new concept of nature, a new concept of science, in other words, a new philosophy.[41]

In the following chapter I shall isolate for special consideration and analysis a crucially important part of this mammoth process: viz. the struggle out of and against Aristotelian (impetus) physics and towards the principle of rectilinear inertia.

Notes

1. P.K. Feyerabend, *Against Method* (New Left Books, London, 1975), p.285.
2. D.J. de S. Price, 'Contra-Copernicus: a critical re-estimation of the mathematical planetary theory of Ptolemy, Copernicus and Kepler', in M. Clagett (ed), *Critical Problems in the History of Science* (University of Wisconsin Press, Madison, 1959), p.198.
3. T.S. Kuhn, *The Copernican Revolution* (Harvard University Press, Cambridge, Mass., 1957), p.154.
4. N. Copernicus, quoted in T.S. Kuhn, *op.cit.* p.138.
5. N. Copernicus, quoted in T.S. Kuhn, *op.cit.* p.141.
6. T.S. Kuhn, *op.cit.* p.171. This aspect of the Copernican Revolution has also been stressed by B. Easlea, *Liberation and the Aims of Science* (Chatto & Windus, Sussex University Press, London, 1973), Chapter 2.
7. R. Edgley writes: 'Connectedness and system are requirements of theory, which seeks unity in the phenomena. It may also be an aesthetic principle, but it's not aesthetic as against scientific (unless the scientific is wrongly identified with the observational, a typically anti-theoretical position)' (private communication).
8. In Chapter 3 I explored the disastrous consequences in Lakatos' work ensuing from his evasion of the question of the earth's motion.
9. D.J. de S. Price, *op.cit.* p.209.
10. T.S. Kuhn, *op.cit.* p.182.
11. See, for example, P. Duhem, *To Save the Phenomena* (University of Chicago Press, Chicago, 1969), Chapter 1.
12. Bellarmine, quoted in S. Drake (trans), *Discoveries and Opinions of Galileo* (Doubleday Anchor, New York, 1957), p.163.

13 Galileo Galilei, in S. Drake (trans), *op.cit.* p.97.
14 Galileo Galilei, quoted in S. Drake (trans), *op.cit.* p.169.
15 Galileo Galilei, in S. Drake (trans), *Dialogue Concerning the Two Chief World Systems* (University of California Press, Berkeley, Second edition, 1967), pp.356, 130.
16 Galileo, in S. Drake (trans), *Dialogue*, *op.cit.* p.272.
17 *Ibid.* p.9.
18 R. Edgley writes: 'As objects of choice conceptual systems are peculiar in this respect, that they are realities that represent (or misrepresent) other objects, realities. In choosing a conceptual system one chooses a certain conception of reality. In choosing the Aristotelian conceptual system you not only exclude the choice of Copernicus' astronomy, you commit yourself to the view that the '*reality*' represented by Copernicus is *impossible*: you assert, in effect, that it is impossible for the Earth to move; and thus false to claim that the Earth moves; and thus wrong to claim that. You commit yourself to *critical opposition* to Copernicus' (private communication).
19 Galileo, in S. Drake (trans), *Dialogue, op.cit.* p.9.
20 *Ibid.* p.9.
21 A. Koyré, *Galileo Studies*, J. Mepham (trans), (Harvester Press, Hassocks, 1978), p.5.
22 When not quoting from Galileo, the discussion of Aristotelian thought which ensues is based almost entirely on Clavelin's rational reconstruction of it — see M. Clavelin, *The Natural Philosophy of Galileo*, A.J. Pomerans (trans), (The MIT Press, Cambridge, Mass., and London, 1974), especially Chapter 1.
23 *Ibid.* p.7.
24 Galileo, in S. Drake (trans), *Dialogue, op.cit.* p.15.
25 Aristotle, quoted by M. Clavelin, *op.cit.* p.19.
26 M. Clavelin, *op.cit.* p.21. The full significance of this remark will emerge more clearly in Chapter 6.
27 Galileo, in S. Drake (trans), *Dialogue*, *op.cit.* p.106.
28 *Ibid.* p.18.
29 *Ibid.* p.40.
30 M. Clavelin, *op.cit.* pp.183–4.
31 Aristotle, quoted in M. Clavelin, *op.cit.* p.29.
32 A. Koyré, *op.cit.* p.4.
33 T.S. Kuhn, *op.cit.* p.95
34 Galileo, in S. Drake (trans), *Dialogue, op.cit.* pp.32–3.
35 *Ibid.* p.139.
36 *Ibid.* pp.140–1.
37 Two recent books dealing with realism are R. Bhaskar, *A Realist Theory of Science* (Harvester Press, Hassocks, 1978), and R. Keat and J. Urry, *Social Theory as Science* (Routledge & Kegan Paul, London, 1975).
38 R. Bhaskar, *op.cit.* p.47.
39 R. Keat and J. Urry, *op.cit.* p.30.
40 Galileo, in S. Drake (trans), *Dialogue, op.cit.* p.133.
41 A. Koyré, *Metaphysics and Measurement* (Chapman & Hall, London, 1968), p.3.

6 IMPETUS, INERTIA AND DISCONTINUITY

6.1 Continuity and Discontinuity

THE Aristotelian view that the (circular) motion of the Earth was impossible in principle was embedded within a network of interconnected assumptions and arguments from which it could not be dislodged without dismantling the entire structure. Some of the most essential stays which supported it were provided by the Peripatetic theory of motion. According to the Peripatetics, the only intelligible natural movement of the earthy, a corruptible heavy element, was directly downwards along a radius of the finite, spherical universe to its centre, the *terminus ad quem* of the motion. Accordingly, to establish the Copernican hypothesis it was not simply necessary to show that it was compatible with astronomical phenomena; it was also necessary to develop a science of motion which could accommodate the diurnal and annual movements of the Earth. It was Galileo who laid the foundations for the new dynamics which was to reach fruition in Newton's *Mathematical Principles of Natural Philosophy*.

Galileo was convinced that his dynamics was a 'very new science dealing with a very ancient subject'. The theories of free fall and projectile motion which he developed in his *Dialogues Concerning Two New Sciences* had, he said, 'opened up to this vast and most excellent science, of which my work is merely the beginning, ways and means by which other minds more acute than mine will explore its remote corners'.[1] The discussion on the Third and Fourth Days of that Dialogue met an undertaking which he had made in his earlier work on the Two Chief World Systems. Having stated the law of free fall — that the spaces passed over by a body starting from rest are to each other as the squares of the times as measured from the moment of release — he is asked whether he has a mathematical proof for this 'remarkable' assertion. Galileo replies through Salviati

SALV. Most purely mathematical, and not only of this, but of many other beautiful properties belonging to natural motions and to projectiles also, all of which have been discovered and proved by our friend. I have seen and studied them all, to my very great delight and amazement, seeing a whole new science arise around a subject on which hundreds of volumes have been written; yet not a single one of the infinite admirable conclusions within this science had been observed and understood by anyone before our friend.[2]

Unwilling to be deflected from his main line of argument, however, Galileo undertook to demonstrate these theorems on another occasion. His *Dialogues Concerning Two New Sciences* fulfilled that promise.

Galileo's heady confidence that he was beginning afresh, and his conviction that he laid the basis for an entirely new science of motion has not gone unchallenged. Speaking generally about the achievements of those who saw themselves as revolutionaries in the sixteenth and seventeenth centuries, Randall warns historians not to accept uncritically their evaluation 'of their own turning away from the heritage of the past. Their consciousness of fresh discovery and radical reorientation obscured the countless bonds of continuity, in materials, methods, and even achievements, uniting them to their predecessors in the late middle ages'.[3]

The view that an unbroken thread linked Galileo's mechanics to earlier studies of motion is particularly associated with the name of Pierre Duhem. On the basis of his study of medieval manuscripts Duhem came round to the view that many of Galileo's ideas had been anticipated in the fourteenth century by a group of scholars located at the University of Paris. In particular, he suggested that Galileo's inertial theory was a more or less logical outgrowth of Jean Buridan's 'impetus' theory. Buridan introduced the impetus concept to account for the persistence of the motions of heavy bodies after their loss of contact with a mover. He then extended it to other related contexts, for example, the acceleration of heavy bodies in free fall, and the eternal persistence of circular celestial motion. Analyzing his arguments, Duhem concluded that 'Galileo's mechanics was the adult form of a living science of which Buridan's mechanics was the larva'.[4]

Another renowned scholar of medieval science, Anneliese Maier, came to a somewhat different conclusion. The differences between impetus theory and inertial physics are so profound, she said, that 'the new thought could develop not *out* of the old, but only *against* it'.[5] Drake concurs — though not out of any genuine conviction but because he sees Maier as having settled an irritating dispute 'related rather to the theory of history than to the ostensible subject matter'.[6] Drake frequently flirts with discontinuity, but his crudely empiricist conscience never allows him to be seduced by it. Both the continuous and discontinuous approaches, he says, are 'potentially fruitful'; one's choice depends 'more on philosophical prepossessions than it does on the record of ideas and events'.[7]

By contrast, Clavelin and Koyré have no hesitation in rejecting Duhem's position. Clavelin dismisses Duhem's view of the relationship between Buridan and Galileo quoted above as being quite untenable. Koyré regards the Scientific Revolution to have involved a 'mutation' in human thought — a term he borrows from Bachelard. He insists that

> in spite of appearances to the contrary, appearances of historical continuity to which Caverni and Duhem give such emphasis — classical physics, issuing from the thought of Bruno, Galileo and Descartes, was not in fact continuous with the medieval physics of 'the Parisian precursors of Galileo': it was from the very beginning located on a different terrain, a terrain that we would like to define as Archimedean.[8]

Continuity or discontinuity: that is the question. Or is it? For on the one hand one simply cannot deny that there are distinct similarities and parallels between Galileo's ideas and those of his Aristotelian and medieval predecessors whom he studied as a young man. On the other hand, there are also profound differences, as Galileo himself implied when he insisted that he was producing a 'very new science dealing with a very ancient subject'. This tension is quite explicit in another assessment of Duhem's position which Koyré made a few years after the one I have just quoted. Koyré writes

> the apparent continuity in the development of medieval and modern physics (a continuity so emphatically stressed by Caverni and Duhem) is an illusion. It is true, of course, that an unbroken tradition leads from the works of the Parisian Nominalists to those of Benedetti, Bruno, Galileo and Descartes . . . . Still the conclusion drawn therefrom by Duhem is a delusion: a well-prepared revolution is nevertheless a revolution, and in spite of the fact that Galileo himself in his youth (as well as at times Descartes) shared the views and taught the theories of the medieval critics of Aristotle, modern

science, the science born from his efforts and discoveries, *does not* follow the inspiration of the 'Parisian forerunners of Galileo'; it places itself at once on a quite different level — on a level which I should like to call the Archimedean one.[9]

Here, in contrast to the previous statement, Koyré concedes that Galileo's thought can be located in an 'unbroken tradition' which leads from Buridan's Parisian school through Benedetti and Bruno to Descartes. Yet, although he recognizes this, Koyré also unhesitatingly affirms that Galileo's achievement was revolutionary and that the transition to his physics was discontinuous. For Koyré, at least, continuity and discontinuity are not mutually exclusive categories; but they are not equally fruitful ways of understanding the historical process either.

My aim in this chapter is to defend a conception of discontinuity which, like that of Kuhn and Feyerabend, sees such transitions as involving the *complete replacement* of one system of thought by another, but differs from theirs in that it emphasizes that such transitions *take time* and are rationally explicable. The raw material for my argument is drawn from the Scientific Revolution (in physical thought), and from Galileo's contribution to it in particular. To this end I shall first describe Buridan's impetus theory, and Newton's inertial theory, and show how they are related to the Aristotelian system of thought. I shall then discuss, and reject two accounts of how the transition took place: one a continuist and the other a discontinuist account. Thereafter I shall describe some of the important steps which Galileo took in founding his new science of motion. I will show how, even while transforming the tradition which he had inherited, he remained ensnared within its confines and so was never quite able to formulate the principle of rectilinear inertia. In the course of the analysis some more of the basic concepts required to develop an adequate theory of discontinuous change will be brought into play. These will be drawn together in the concluding section which summarizes the argument for the position which I am defending in this study.

6.2 Impetus Theory

The priority which Aristotle granted to rest over motion

demanded that he identify the cause or causes of which motion was the effect. Rest, being the natural condition of (terrestrial) bodies, could persist of its own accord; motion could not. In the case of natural motion the pressure to actualize unrealized possibilities drove terrestrial bodies to their natural places. The temporary persistence in motion of 'violently' moved bodies was less easily accounted for. Aristotle spells out the difficulty when he says that 'If it is true that everything that is moved, except things which are self-moved, is moved by something, how comes it that some things are moved continuously, though that which has caused them to move is no longer in contact with them, as, for instance, things *thrown* ?'[10]

Two theories were put forward by Aristotle to explain the continuing movement of a projectile once it had lost contact with the thrower. Both relied on the air to serve as a propellant. The first, the so-called antiperistasis theory, asserted that as the body moves forward the air which it displaces rushes in behind it, and propels it further. Aristotle rejected this view in favour of a more complex alternative. This second theory postulated that the original motor not only sets the projectile going, but also moves the air next to it, while simultaneously transmitting to the air the power to act as a motor. This power passes from one portion of air to the next as the body moves through space; by virtue of it the air is able to sustain the motion of the projectile.

These theories were already criticized by John Philoponus in the sixth century; by the fourteenth century Aristotle's explanation of projectile motion had emerged as one of the major weaknesses in his dynamics. Buridan rejected both the antiperistasis theory and Aristotle's preferred alternative. He noted, for example, that a ship continued to move long after the haulers had stopped pulling it. Its persistence in motion could not be explained by arguing that displaced air rushed in and pushed it from behind. A sailor on deck experienced the resistance of air from the front; he did not feel a breeze from behind pushing him. The alternative theory was equally defective, though. If the ship were covered with a cloth, said Buridan, and the cloth along with the ambient air were removed it would not stop on that account. Yet is should on Aristotle's theory, according to which the power to sustain motion was located in the surrounding medium.

On the basis of these and other equally ingenious arguments Buridan rejected the view that the persistence of violent motion demanded the continuous action of an external agent. In particular, the air appears to retard motion, not to perpetuate it. Therefore, wrote Buridan

> it seems to me that it ought to be said that the motor in moving a moving body impresses in it a certain impetus or a certain motive force of the moving body, [which impetus acts] in the direction toward which the mover was moving the moving body, either up or down, or laterally, or circularly. And by the amount the motor moves that moving body more swiftly, by the same amount it will impress in it a stronger impetus. It is by that impetus that the stone is moved after the projector ceases to move. But that impetus is continually decreased by the resisting air and by the gravity of the stone, which inclines it in a direction contrary to that in which the impetus was naturally predisposed to move it. Thus the movement of the stone continually becomes slower, and finally that impetus is so diminished or corrupted that the gravity of the stone wins out over it and moves the stone down to its natural place.[11]

Having thus introduced his impetus concept, Buridan went on to develop the idea further. One objection that he had raised to Aristotle's theory of projectile motion was that, if the air sustained motion rather than resisting it, as the Stagirite held, a feather should be projected further than a stone. It is not, of course, and Buridan explained why

> By the amount more there is of matter, by that amount can the body receive more of that impetus and more intensely. Now in a dense and heavy body, other things being equal, there is more of prime matter than in a rare and light one. Hence a dense and heavy body receives more of that impetus and more intensely, just as iron can receive more calidity than wood or water of the same quantity.[12]

Buridan also invoked his impetus theory to account for the acceleration of heavy bodies in free fall. Initially, he said, the body falls downward solely under the influence of its gravity. This movement impresses in it an impetus, which acts in conjunction with the gravity to move it a little faster. The faster it goes the more intense is the impetus impressed on it; thus, the further away it is from its *terminus a quo* the faster it will be travelling.

Unlike Oresme, for example, Buridan did not regard

impetus to be self-expending. It was permanent (i.e., it did not exhaust itself of its own accord but only in overcoming air resistance and gravity). This enabled him to use his impetus theory to explain the perpetual movement of celestial bodies. As I pointed out in the previous chapter, Aristotle regarded these to be composed of a 'quintessence', neither light nor heavy, which moved naturally in a circle for ever, sustained by eternal and immutable Motor Intelligences. Infusing Christian elements into this view, Buridan suggested that God had initially impressed an impetus on the celestial orbs when he created the world, and that 'these impetuses which He impressed in the celestial bodies were not decreased nor corrupted afterwards, because there was no inclination of the celestial bodies for other movements. Nor was there resistance which would be corruptive or repressive of that impetus'.[13] In other words, since celestial bodies were not endowed with gravity and moved freely through the superlunar realm, once God had set them in motion, the permanent impetus which he impressed on them would ensure that they moved perpetually. With that, Aristotle's Motor Intelligences were rendered redundant.

It is plain that Buridan's impetus concept was a substantial innovation, and a great improvement on Aristotle's theory of projectile motion. However, it must be stressed that Buridan was always careful to distinguish the impetus as a *cause*, from motion, which was its *effect*. Typically, he writes, 'since every motion arises from a motor being present and existing simultaneously with that which is moved, if the impetus were the motion, it would be necessary to assign some other motor from which that motion would arise. And the principal difficulty would return. Hence there would be no gain in positing such an impetus'.[14]

This means that celestial bodies, say, move perpetually in circles not because their motion is conserved, but because the permanent impetus which sustains that motion is not 'diminished or corrupted' in any way. It is the cause of the motion that is conserved, not the motion itself. The impetus concept was developed within the framework of Aristotelian cosmology. Basically Buridan did not reject Aristotle's claim that 'everything that is moved, except things which are self-moved, is moved by something'. What he did was to place the cause of motion in the body itself, unlike the Stagirite, who located it in the medium. The significance of this point will emerge shortly when I discuss the relationship between impetus theory and inertial physics.

6.3 Inertial Physics

Reminiscing as an old man about his achievements, Newton described to a younger friend and admirer, William Stukeley, how he had hit upon his universal law of gravitation. Many years later Stukeley recalled the incident:

> After dinner, the weather being warm [the date was 15 April 1726] we went into the garden and drank thea, under the shade of some appletrees, only he and myself. Amidst other discourse, he told me, he was just in the same situation, as when formerly, the notion of gravitation came into his mind. It was occasion'd by the fall of an apple, as he sat in a contemplative mood. Why should that apple always descend perpendicularly to the ground, thought he to himself. Why should it not go sideways or upwards but constantly to the earth's centre ?[15]

Let us pause here a moment. To the Aristotelians, these questions were easily answered. The apple, they would say, falls to the ground because, as a heavy terrestrial body, it has a natural tendency to move downwards along a radius of the spherical universe to its centre, at which is located the Earth. It cannot move upwards because it is a heavy body, and to be heavy means to go downwards. It can only move sideways 'unnaturally' (i.e., if propelled by an externally acting agent which violates its inherent tendency to return to a state of rest). No such agency is at work here; as Stukeley says, Newton saw the apple fall freely to the ground. As such, for the Aristotelians, its movement was natural, not violent, and could not be anything but rectilinear: it was impossible for a heavy body released from on high to do otherwise.

The Aristotelians, then, would have been quite capable of providing Newton with an explanation of the apple's behaviour. But Stukeley not only makes no reference to the account of this event that they would have given; he presents Newton's view as if it were the only possible account of it. The apple, he says, falls to the Earth's centre, and

> Assuredly, the reason is, that the earth draws it. There must be a drawing

power in matter: and the sum of the drawing power must be in the earth's center, not in any side of the earth. Therefore does this apple fall perpendicularly, or towards the center. If matter thus draws matter, it must be in proportion of its quantity. Therefore the apple draws the earth, as well as the earth draws the apple. That there is a power, like that we here call gravity, which extends itself thro the universe.[16]

Stukeley wrote this in 1752, 110 years after Galileo had died and Newton was born. By that time it was no longer necessary to recognize that there could be, or had been, an alternative dynamics. Aristotelianism had been effectively routed, and inertial or classical physics had established itself. The rectilinear movements of falling apples and the elliptical paths of planets around the Sun were seen to be governed by the same laws. Universal gravitation finally dissolved the Aristotelian schism between the terrestrial and the celestial; in fact, as Koyré puts it, 'all the phenomena which for centuries baffled the sagacity of astronomers and physicists (such, for instance, as tides) appear as a result of the concatenation and combination of the same fundamental laws'.[17]

'If I have seen further than other men', said Newton, 'it is because I have stood on the shoulders of giants'. And indeed the synthesizing power of his gravitational concept presupposed, and was made possible by, even more fundamental developments initiated by his predecessors. In particular, it was they who laid down what it was that a science of dynamics needed to explain. To the Aristotelians the perpendicular fall of an apple was part of the natural course of events. It was the kind of thing one expected apples to do, and it was not necessary to appeal to external forces to account for it.[18] The people who developed the science of mechanics in the seventeenth century thought otherwise. After a long and tortuous journey, some stages of which will be discussed later, they insisted that an external force was needed to explain free fall, and indeed to explain the movement of any body which was not behaving in accordance with the inertial principle — and that meant any real, material body whatsoever.

In 1644 Descartes first explicitly formulated the principle of inertia. As stated by Newton, who mistakenly attributed it to Galileo, this principle asserts that 'Every body perseveres in its state of rest, or uniform motion in a right line, unless it is compelled to change that state by forces impressed thereon'.[19]

For all its apparent simplicity, this law embodies a profound revision in the concepts of motion and of the nature of matter. It is this aspect of the Scientific Revolution which has been forcefully brought to light by Koyré,[20] and which has led him to insist that it was, in effect, a discontinuous transition. The fundamental difference between inertial or classical physics and Aristotelian physics, he argues, lies in the quite distinct ways in which they thought about motion. Descartes, as a matter of fact, said that he had no idea what the Peripatetics meant when they defined motion as the actualization of the potential insofar as it was potential (i.e., insofar as a particular substance was lacking in it). According to Koyré, this is because he, like ourselves, regarded motion to be a *state* akin to rest, and not a *process*, as it was for the Aristotelians.

For the Aristotelians the study of motion was part of a more general science of change: change of place was one of three processes whereby an existing substance lacking a particular property could acquire it. (Natural) motion had an ontological function: it involved the actualization of potentialities, experienced as a privation, and it ceased when a (terrestrial) body reached its natural place. By moving, bodies changed their conditions of existence; correlatively the directions in which they spontaneously moved to restore cosmic order were dictated by their natures.

An entirely different conception of matter and of motion is implied by regarding the latter as a state, on the same ontological footing as rest. For what this indicates is that a body is *indifferent* as to whether it is in a state of motion or of rest. On the one hand, matter is rendered *inert*. A body does not fulfil itself through movement; it remains passively at rest or in a state of (uniform rectilinear) motion until disturbed by an external force. On the other hand, relieved of its ontological function, motion becomes a phenomenon *sui generis*. Like rest, uniform rectilinear motion persists of its own accord. It is not an effect, as it was in impetus theory, and unless and until it changes either in magnitude or in direction, no cause need be invoked to explain it. As a state (uniform rectilinear) motion is as indestructable and as permanent as rest. As a process motion was transitory; it persisted only so long as it was sustained by privation (natural motion) or an external agent (Aristotelian violent motion).

When I discussed the tower argument in Section 5.3 I quoted Salviati as saying that for the Peripatetics it was impossible for a rock released from on high to fall freely with a mixed straight and circular motion. The reason given was that, being a heavy body, the stone moves directly and vertically to the Earth: this is the only spontaneous movement compatible with its nature. This constraint is now lifted. Since motion no longer expresses the nature of the moving body, the latter cannot influence the permissible modes of combination of two or more motions in the same body. For the Aristotelians, different movements in the same body should interfere with or impede each other (as in Buridan's impetus theory), or could even be incompatible with each other (as in the case of the tower argument). In classical physics, by contrast, a body may be endowed with any number of motions. These can be combined using geometrical rules to produce a resultant, and a given motion can be broken down using similar procedures into its several components. Here, then, motion is no longer subjected to the physical nature of the bodies that endure it.

Treating motion as a state ontologically equivalent to rest, and not as a transitory process, has another important consequence. The Aristotelians insisted that the transition from motion to rest involved a change in the conditions of existence of the body. In classical physics it involves a change of state which, while requiring the application of a force, in no way affects the body itself. The fact that the body is at rest or in motion is irrelevant from its point of view, so to speak. This implies, says Koyré, that we cannot ascribe motion to a body in isolation.[21] The state of rest or of motion can only be attributed to it in relation to another body assumed to be in motion or at rest, and it is quite arbitrary which state is attributed to which body. Motion, in other words, is not simply a state; it is a *relational* state.

The transitoriness of terrestrial motion in the Aristotelian cosmos was indicative of their treatment of it as a species of becoming, not of being — as *teleological*. As such they deemed it to be unamenable to mathematical analysis — Aristotle himself had said that there is no change in numbers and in figures. By transforming it into a state that, in certain circumstances, was capable of indefinite self-perpetuation, classical physics raised it to the level of being where it could be analyzed mathematically. The motion with which it deals is that of abstract bodies moving perpetually in an idealized realm. Free from the disturbing influences of external forces they travel forever in a straight line in the void of Euclidean space. Aristotelian bodies, by contrast, are generable and corruptible, subject to qualitative change and decay; they are the bodies of everyday sense experience. They exist in the fulness of their being only when at rest in their natural places. Their behaviour is not determined by mathematical laws, like that of (what Koyré calls) the Archimedean bodies of Galileo's dynamics. It is dictated by cosmological imperatives which compel them spontaneously to restore order whenever it is disrupted.

It is these considerations which have led Koyré to stress repeatedly that the Scientific Revolution involved the destruction of the cosmos and the geometrization of space. The latter amounted to the replacement of a finite spherical and hierarchical universe, with its precisely specified directions and places, by the infinite, abstract, and directionless Euclidean void. The disappearance of the cosmos ultimately entailed the construction of a universe in which the separation between the terrestrial and the celestial was dissolved, and the realms of becoming and of being were placed on the same ontological level. Underpinning these changes was the articulation of that concept of motion and that view of matter which I have discussed in this section.

6.4 From Impetus to Inertia: Shapere's Continuism

There are several superficial similarities between impetus theory and some of the basic concepts of classical mechanics. For example, in Section 6.2 I quoted Buridan as saying that the amount of impressed impetus will be determined by 'the amount the motor moves that moving body more swiftly' (i.e., by the transferred speed) and by the quantity of 'prime matter' which it contains. This led Duhem and, more recently, Clagett, to suggest that Buridan's concept of impetus is similar to Newton's concept of momentum, or quantity of motion, the product of the body's mass and its velocity. Clagett concedes that momentum and impetus are ontologically different, since the latter is a kind of force, a cause of motion, but nevertheless remarks that 'while the affirmed ontology of impetus would seem to differentiate it from the later concepts, yet the terms of its *measure* as presented by Buridan make an analogue with momentum, i.e., this quality which is motive force for Buridan turns out to be described in dimensions analogous to those of Newton's momentum'.[22] Consistent with this interpretation, Clagett also remarks that Buridan's concept of 'quantity of prime matter' is 'a kind of analogue of the "mass" of early modern physics'.[23]

Another point of resemblance between the impetus concept and the inertial concept that Clagett detects concerns Buridan's use of the former to explain the perpetual motion of celestial bodies. 'It is here', writes Clagett, 'that Buridan comes closest to producing an analogue with the later inertial idea'.[24] The reason is that since Buridan's impetus is permanent, it is conserved in the absence of air resistance and the natural downward tendency of terrestrial bodies. Accordingly, if the impetus in question were rectilinear rather than circular, a body in which it is impressed would continue to move in uniform rectilinear motion forever under its influence.

It is parallels like this which have led Shapere to suggest that impetus theory 'provides an excellent case study to illustrate the comparability of scientific concepts and the continuity of their evolution from one "tradition" to another'.[25] The impetus concept, he goes on to say, straddles the divide between the Aristotelian tradition and inertial theory, with which it has a large number of 'affinities', 'resemblances and continuities'. While not yet inertia nor momentum, it is 'a visible move away from fundamental Aristotelian conceptions . . . and in the definite direction of classical mechanics'.[26]

Shapere isolates for special consideration three steps along the gentle slope that allegedly leads, or led, from impetus to inertia.[27] These are:

1 The conception of the possibility of motion in a void (i.e., of the possibility of motion free of all external forces);
2 The abandonment of the idea that there is circular as well as rectilinear impetus; and

3 The abandonment of the idea of impetus as an internal force distinct from the body and maintaining it at a constant speed.

With these modifications, he alleges, Aristotelian physics could have been transformed into inertial physics. I shall discuss each of them in turn.

The first obstacle in the way of the inertial principle which Shapere isolates is the rejection of the view that motion in a void was impossible. Shapere locates the difficulty in what has been called the fundamental law of Peripatetic dynamics, viz., that the velocity of a moving body $v=F/R$, where F is the absolute weight of the body (in natural motion) or the propelling force (in violent motion), and R is the resistance of the medium. Plainly in a vacuum R is zero, and so the velocity in a void would approach the infinite. Aristotle deemed this to be impossible.

Having identified the obstacle in this way, Shapere regards it as being circumvented by the introduction of expressions for the velocity in which v is not indeterminate when R is zero. And he notes that in the Middle Ages one such expression was put forward, viz. the velocity was taken to be proportional to $(F-R)$. Galileo adopted this formulation for natural motion in one of his early works; by this simple expedient, it seems, motion in a void became a physical possibility.

Shapere's analysis is grossly misleading. Far more was required to render motion in a void physically plausible than the development of an appropriate formula for the velocity. In the first place motion in a void was excluded by Aristotle's cosmological premises. In a void there could be neither up nor down — by definition it lacked such distinctions. But natural terrestrial motion is unintelligible without termini; in a void, therefore, a body would have no reason for moving in one direction rather than another, and so would have no reason to move at all. And in the second place Aristotle contended that motion could only be continuous if it was resisted. The medium, though opposed to the motion, 'preserved its physical reality from one moment to the next, and this precisely by forcing the moving body to overcome the opposing force'.[28] As Clavelin notes, this view was yet another manifestation of Aristotle's view of motion as transitory. As a state, rest could persist of its own accord.

Motion could not: a resisting medium was a *sine qua non* of its very existence. In other words, lying behind the formula $v=F/R$, which is actually Duhem's and not Aristotle's anyway, lies a set of interconnected arguments and considerations which could not have been overlooked by anyone who wished seriously to defend the view that motion in a void was plausible. To render it more than an abstract possibility it was necessary to undermine the foundations of Aristotelianism itself.

Similar considerations apply to the second of the three obstacles which Shapere identified on the path from impetus to inertia: viz. the jettisoning of the idea that there was circular as well as rectilinear impetus. Shapere suggests that this was a particularly difficult impediment to dislodge because of the fundamental role which circular motion played in Greek and medieval science. And he notes that Galileo opted for circular motion as natural in both the sublunar and the celestial domains because he believed that only circular motion was appropriate to an ordered, perfect world. According to Shapere, this indicates that 'in this respect, at least, far from rejecting all of classical thought, Galileo claims to be carrying out more consistently some of its deepest assumptions'.[29]

In Section 6.6 I shall support Shapere's general interpretation of Galileo's achievement and, in particular, I shall follow him, Koyré and Clavelin in arguing that Galileo did not put forward the principle of rectilinear inertia. Shapere himself explains why when he writes 'the idea that any body might continue in uniform rectilinear motion forever is — for Salviati as it was for Aristotle — senseless. For Galileo to have asserted the principle of inertia would have contradicted fundamental principles which, at least in this portion of his writings, he accepted from the Aristotelian tradition'.[30] Now if the principle of inertia was 'senseless' because it would have contradicted certain 'fundamental principles' of Aristotelian thought it is plain that the transition to it must have required the rejection of those principles themselves. Shapere has identified one of these: the conception of the universe as ordered. I shall stress another: the view that gravity is an inherent property of bodies. Galileo did not abandon these ideas. But he did reject the fundamental Aristotelian dichotomy between the celestial and the terrestrial and for that reason was able to regard the same kind of motion as natural to both. This was a crucially important step on the path to the principle of inertia: in conjunction with the destruction of the cosmos, it ultimately led to the annihilation of Aristotelianism and impetus theory. In other words, although the transition from Peripatetic mechanics to classical physics was neither achieved by a single person, nor in a flash, it was discontinuous in the sense that it involved the complete replacement of one structure of thought by another. The fact that it took time to dismantle Aristotelianism and to purge its residues from thought does not mean that the change was continuous.

The third barrier between the impetus theory and the principle of inertia which Shapere isolated is the idea that impetus was an integral force distinct from the body and maintaining it at constant speed. It is obvious that this notion could not be rejected without abandoning Aristotelianism itself. In Section 6.2 I stressed that Buridan insisted that impetus was the cause of which motion was the effect, and I noted that he was careful to distinguish between them. This was because he introduced the impetus concept precisely to explain why (violent) motion persisted after the projectile had lost contact with the mover. This is not a problem for inertial physics; on the contrary, it is the *failure* to persist in (uniform rectilinear) motion which it seeks to explain. Impetus theory, in other words, was a response to Aristotle's view of motion as transitory. It was formulated on the assumption that (sublunary) matter, far from being indifferent to motion and rest, had a spontaneous tendency to return to rest and, because of its inherent gravity, resisted motion. The only way to transfer to motion itself the properties which Buridan attributed to impetus was by demolishing the structure of thought within which the impetus concept was embedded.

If my argument is correct, it follows that Shapere has been completely misled by the superficial resemblances between the meanings of concepts like impetus and inertia. Accordingly he has imposed a spurious continuity on a fundamentally revolutionary transformation, blurring differences and dissolving specificities. He has failed properly to locate these cardinal ideas in the conceptual frameworks which supported them. As a result he does not appreciate that to progress people like Galileo had to burrow deep into the recesses of Aristotelian thought and to undermine it from within. The seeds of a new science of motion were planted in the spaces which he thereby created. If he had been content merely to scratch at the surface of the Peripatetic system, timidly emphasizing similarities and continuities, the Scientific Revolution as we know it would never have taken place.

Shapere's continuism derives what plausibility it has from the stress which he (and Clagett) place on the similarities in the meanings of concepts and ideas developed in Buridan's physics and in classical mechanics. This approach, however, has one fatal flaw: continuism abstracts these concepts from the historical context in which they were used by living speakers and writers. It is only with the wisdom of hindsight, and standing outside the historical process, that we can detect the similarities — and differences — between impetus theory and inertial physics. Of these Buridan was quite oblivious, of course; when he put forward impetus theory he knew nothing of Newton's inertial theory. Even to its protagonists, impetus theory was, at least in part, 'a visible move away from fundamental Aristotelian conceptions (e.g., it allows the sustaining force in projectile motion to be internal and incorporeal)',[31] as Shapere puts it. But *for them* it was not *also* a visible move 'in the definite direction of classical mechanics', as he goes on to say.

By abstracting concepts and ideas from the statements in which they were used Shapere and Clagett have fallen into a trap. They have mistakenly identified the history of scientific theories with the history of *disembodied* systems of thought. It is, as Dunn notes, a 'bloodless' history, dealing in 'reified abstractions'

> All the statements contained in it are statements about the relationships of propositions to propositions. Men, breathing, excreting, hating, mocking, never step inside it. Their role is merely to label a particular set of propositions with the name which they bear themselves. Their names appear in this story but never their selves. It is a tale to be told by clever and subtle men, and it signifies much but in it there is neither sound nor fury.[32]

By failing to appreciate that the formulation of a scientific theory is a social *activity* performed by conscious human agents, this approach to history misses 'the *point* of the original intellectual enterprise'.[33] To establish that point in the case of Buridan's impetus theory is to ask what *his reasons* were (or what *the reasons* were) for advancing the impetus concept. It is to establish what Dray[34] has called the rationale of what Buridan did; it is to explain why he found it necessary to introduce the new theory, to detect the problems which he took to be pressing and in need of solution. In other words, if we are to grasp adequately the significance of Buridan's impetus theory we must locate it within the framework of his fundamental presuppositions, those 'blocking-off points to the account which can be given of his argument',[35] as Dunn calls them. This is the procedure that Koyré and Clavelin have adopted and on which I have relied above.

Of course, one reason that Buridan could *not* have had for developing impetus theory was to bridge the gulf between Aristotelian dynamics and classical mechanics. This point is lost if we analyze the development of scientific thought by concentrating solely on superficial similarities between the meanings of words. More generally, the aims and objectives of conscious human beings are effectively dismissed as irrelevant to an understanding of the historical process. That process accordingly assumes a life of its own. It is driven forward by the motor of its own inner logic, passing steadily and 'reasonably' from one phase to another, as the implications of each successive advance are worked out. As Drake notes, in this case, we are left with the overall impression that 'medieval impetus theory led on logically and almost inexorably to inertial concepts in the works of Galileo, Descartes and Newton'.[36] Continuist history, in other words, goes hand in glove with a subjectless epistemology of the type so beloved by Popper.[37]

Although impetus theory did not lead 'logically and inexorably' on to inertial physics, it did play an important role in creating some of the material conditions necessary for the ultimate rise of classical mechanics. For example, I pointed out in Section 6.2 that to explain the acceleration of a freely falling heavy body, Buridan suggested that after an initial period in which it was propelled solely by its gravity, a quantum of impetus was added to it. Thereupon it moved faster, and acquired a further, and larger, dose of impetus, accelerating accordingly. Buridan's predecessors had argued that heavy terrestrial bodies falling freely accelerated as they grew closer to the centre of the universe, their natural place, and the *terminus ad quem* of natural rectilinear motion. His impetus theory of free fall, by contrast, suggested that they accelerated as they moved further away from their *terminus a quo* — a point which Galileo only managed to establish with immense difficulty. This made it possible — as Koyré puts it — for Buridan's successors to separate the body's motion from the idea of a goal towards which it moved, and thus contributed to undermining the absolute separation between motion as transitory and rest as permanent.

This idea, the idea that impetus theory 'made it possible to think in certain ways that had previously been precluded or at least discouraged by the Aristotelian system',[38] has also been stressed by Shapere. For example, he notes that the impetus concept drew attention to the quantitative features of motion. I assume that by this he means that Buridan took the impetus to be proportional to the initial speed of, and to the quantity of 'prime matter' in, the body in which it was impressed. It was just the affinity between impetus as measured by this double proportionality and the concept of momentum in classical physics which led Duhem to assert that Buridan's mechanics had anticipated Newton's, and which has inspired continuist interpretations of the transition from impetus to inertia. In the light of the above analysis we can now place these continuist claims in perspective.

From the vantage point of the outcome of an historical process continuism detects anticipations, analogues, precursors of a particular idea or conception. In doing so it forces us to recognize that, as historical agents, people are immersed in the flux of history, that the past seeps into them through socialization, and that even as they rebel against it, they cannot tear themselves wholly free from it. Herein lies a strength of the continuist approach. One of its major weaknesses is that, misled by superficial similarities, it blurs and dissolves differences. Admittedly a concept like impetus made certain kinds of intellectual moves possible. However, these possibilities were doomed to remain purely abstract and speculative as long as the concept remained embedded in its underlying presuppositional framework. Access to that framework is gained through analysing the 'point of the original intellectual enterprise' (Dunn) which generated a notion like impetus in the first place. Within the (perhaps rationally reconstructed) framework thus exposed to view there will be 'blocking-off points', fundamental presuppositions which *limit* the range and scope of the articulation of the theory in question. Those limits can be tested, can almost be transcended, as when Buridan remarks that a smith's wheel may 'well' continue to move indefinitely in the absence of all resistance.[39] While purely suggestive at this stage of the development of mechanics, comments like this also indicate points at which that mechanics is defective. They disturb the tranquility of an established system of thought, opening up new vistas which do not develop beyond imaginative constructs until a new theory built on new foundations can meaningfully accommodate them. As Drake noted, 'to the extent that impetus theory paved the way for eventual acceptance of the idea that motion might be perpetually conserved, it rested on a philosophical basis that inhibited the taking of that step'.[40] In the inability to recognize this point lies one major flaw in the continuist account of the Scientific Revolution, an inadequacy which it conceals by describing the transition from impetus to inertia as 'inexorable' and 'logical'.

6.5 From Impetus to Inertia: Discontinuity and Incommensurability

In the opening pages of this book I pointed out that Foucault and Koyré believed that discontinuous transitions had occurred in history. Both implied that there had been drastic transformations of the human understanding in which one conceptual framework was rejected and replaced by another. The earlier theory did not live on as a limiting case of its successor. Rather it was eliminated root and branch and a quite different theory was produced in the process. I suggested at the time that the two theories may be said to be 'incommensurable' with each other.

My use of this term reflects my sympathy with some, but by no means all, features of the controversial doctrine of incommensurability. The doctrine has been defended by Kuhn and, independently of him, by Feyerabend. One of Kuhn's motives for introducing it was his belief that although

Aristotelian thought was different in many respects from modern physical theory, it provided an orderly or 'rational' account of a great variety of phenomena.[41] An original source of Feyerabend's arguments for incommensurability was his criticism of the monistic tendencies in the adherents of the Copenhagen Interpretation of the quantum theory, as evinced in particular in the doctrine of meaning invariance (see Section 4.1). Feyerabend does not think that every major development in science involves a transition between incommensurable frameworks; however, he believes that the transition from impetus to inertia was of this type.

According to Kuhn, there are three different ways in which two theories or 'paradigms' (see Chapter 1) may be incommensurable and they are as follows:

1 Adherents of competing paradigms will disagree about the methods, the problem fields and the standards of solution which should apply in their discipline;
2 Changes in paradigm change the meaning of established and familiar concepts which constitute the scientists' vocabulary, so that the conceptual framework which they use to make sense of the world is radically transformed; and
3 In some difficult-to-specify sense, adherents of competing paradigms practice their trades in different worlds; a paradigm, Kuhn suggests rather tentatively, structures our perceptual experience of reality.

It is important to realize that these suggestions form part of Kuhn's general view that there can be discontinuous transitions in history. As such he has generated them in opposition to the historical thesis that science always develops cumulatively, and a related philosophical claim that relations of deducibility hold between successive scientific theories. Concerning the latter I have already quoted Kuhn as saying that

> Even a theory like energy conservation, which today seems a logical superstructure that relates to nature only through independently established theories, did not develop historically without paradigm destruction. Instead, it emerged from a crisis in which an essential ingredient was the incompatibility between Newtonian dynamics and some recently formulated consequences of the caloric theory of heat. Only after the caloric

theory had been rejected could energy conservation become part of science. And only after it had been part of science for some time could it come to seem a theory of a logically higher type, one not in conflict with its predecessors. It is hard to see how new theories could arise without these destructive changes in beliefs about nature. Though logical inclusiveness remains a permissible view of the relation between successive scientific theories, it is a historical implausibility.[42]

Considerations like these have also led Feyerabend to defend the doctrine of incommensurability. He has concentrated on the second of the two ways in which Kuhn has said that theories can be incommensurable, viz. conceptual disparity. In his own words, he initially argued for

the principle that the interpretation of an observation language comes from the theory that explains what we observe, and changes as soon as this theory changes. I realised that the principle might make it impossible to establish deductive relations between rival theories and I tried to find means of comparison that were independent of such relations . . . . Comparison by *content*, or *verisimilitude* was of course out. But there certainly remained other methods.[43]

These other methods of comparison are 'subjective' in that, rather than being based on logic and experience, they are informed by 'aesthetic judgements, judgements of taste, metaphysical prejudices, religious desires'.[44] Though perhaps not 'objective' Feyerabend insists that they are, nevertheless, reasonable in the sense that they reflect the wishes of a sizeable number of researchers.

It is to Kuhn's and Feyerabend's credit that, through the doctrine of incommensurability, they have focussed attention on the importance which far-reaching *conceptual* innovation has played in the growth of science. The rejection of one theory and its replacement by another 'incommensurable' alternative involves something more and other than, say, devising crucial experiments between the two. Although empirical considerations play a role in effecting discontinuous transitions, to replace one theoretical framework with another it is also necessary to forge a new cognitively adequate conceptual apparatus. It is precisely for this reason that I have insisted that (conceptual) revolutions are protracted and laborious processes, and not sudden, more or less unstructured events.

One weakness in the incommensurability thesis as defended by Kuhn lies in his failure to recognize this. As I stressed in Chapter 1, he tends to emphasize the role of the arbitrary and the non-rational in accounting for how discontinuous transitions occur between different frameworks. In appealing to the notion of the gestalt switch, Kuhn compares paradigm shifts to the sudden and comprehensive flips of perception experienced when looking at the 'duck-rabbit' or Necker cube. Thus he writes that 'just because it is a transition between incommensurables, the transition between competing paradigms cannot be made a step at a time, forced by logic and neutral experience. Like the gestalt switch, it must occur all at once (though not necessarily in an instant) or not at all'.[45]

In similar vein, Feyerabend stresses that rebels who want to win support for a view that is incommensurable with prevailing orthodoxy will have to rely on non-rational techniques of persuasion. His conception of how discontinuous transitions occur reflects the greater emphasis which he places on language and meaning in his discussion of incommensurability. He writes

> How is the 'irrationality' of the transition period overcome ? It is overcome in the usual way . . . i.e. by the determined production of nonsense until the material produced is rich enough to permit the rebels to reveal, and everyone else to recognize, new universal principles. (Such revealing need not consist in writing the principles down in the form of clear and precise statements.) Madness turns into sanity provided it is sufficiently rich and sufficiently regular to function as the basis of a new world view.[46]

Both of these 'theories' of how the transition occurs from one system of thought to an incommensurable alternative devalue the role of reasoning or argument as an agent of scientific change. It is suggested that, when all is said and done, the adherents of the old and of the new live in different worlds, talk past each other, and are incapable of rationally criticizing one another. As long as one group does not impose its will on any other, alternative schemes of thought will develop in parallel in perpetuity (Feyerabend). Alternatively the new view prevails, not because its opponents are convinced of its truth, but because they eventually die, and a new generation grows up that is familiar with it (Kuhn).

These drastic — and rather unpalatable — consequences of the doctrine of incommensurability play into the hands of those who would defend a continuist view of historical change. As a matter of fact Shapere's arguments discussed in the previous section were specifically developed as a counter to Kuhn's and Feyerabend's conception of discontinuous transitions, and its relativistic implications. As far as he is concerned, the choice before us is plain: either we adopt discontinuity, and with it the irrationalism of the period of transition, ultimately landing up in a 'relativistic bind', or we adopt continuity, and with it a view of science as proceeding 'by a succession of developments that can be described as reasonable'.[47]

This choice is indeed a stark one. But its starkness is a consequence, at least in part, of the failure to transcend the limits imposed by the dichotomy between the contexts of discovery and of justification. As I have repeatedly stressed, if this dichotomy is adopted as a working premise, albeit not consciously, it immediately locates philosophers outside the historical process. They do not and cannot take a participant's perspective on historical change. Rather, they *presuppose* that certain kinds of major historical change have taken place without seriously considering how they occurred, and they then try to specify ways of choosing (rationally) between the *outcome* of such changes, viz. fully developed, 'ready articulated theories'.

This is precisely what has happened here. The defenders of discontinuity and incommensurability, like Feyerabend and Kuhn, tend to argue that one cannot choose between *comprehensive world views* on the basis of logic and experiment. All that remains are our subjective wishes and personal inclinations. Their critics accuse them of playing fast and loose with meaning, attempt to expose inadequacies and contradictions in their project, and show how it is possible to make 'objective' comparisons between *comprehensive world views*. For both sets of protagonists in the dispute the choice that has to be made is one between richly articulated theories. For Kuhn they are adopted whole in a 'gestalt switch'; for Feyerabend the commitment is informed by inclination and interest. Both they and their opponents take the production of new theories to be psychological and deeply personal acts of individual creativity. They differ only in that whereas one party to the dispute insists that 'objective' criteria can and should be used to criticize and to choose between full-blown theories, the other insists that this is not possible nor, as far as Feyerabend is concerned, is it reasonable or compatible with individual happiness. Either way, logic is played off against psychology, the rational against the irrational, argument against inclination, the gradual against the sudden, the evolutionary against the cataclysmic. These are the landmarks on a shared terrain the topography of which dictates the terms of the engagement between the critics and the defenders of the incommensurability thesis.

Any adequate theory of discontinuous transition must suspend these sterile dichotomies and, in particular, the dichotomy between the contexts of discovery and justification. It is only by doing so that it is possible to take a participant's perspective on the history of science. The central problem in the incommensurability debate is taken to be that of choosing between comprehensive world views, between, say, Aristotelian cosmology plus impetus theory, and inertial physics. This problem was simply not one that was faced by those who travelled from impetus to inertia in the seventeenth century. *They* did not have to choose between two ready made systems of thought; they rejected one and dismantled it, simultaneously constructing another. That process took time and involved an immense emotional and intellectual effort — not a gestalt switch or the production of nonsense.

I have frequently suggested that Aristotelian physics and classical mechanics are incommensurable systems of thought. This reflects my view that — within the confines of the Peripatetic system — it was impossible to accomodate the motion of the Earth and to develop a dynamics that could deal with that idea. A different conception of motion was required — one which regarded it to be a state rather than a process. Against Kuhn and Feyerabend, however, I want to insist that the process whereby the classical conceptual framework was developed was a lengthy one and is not (necessarily) rationally unintelligible. Just how painful and tortuous it was will emerge from the next section, in which I trace Galileo's path to his principle of inertia. If my exposition is correct, it follows that enough of the meaning of certain crucial dynamical terms remained for Galileo to argue his way out of the Aristotelian framework, which he never managed to shed completely anyway. This does not mean that (the contents of) arguments alone are sufficient to effect discontinuous transitions. On the other hand it does suggest, against Kuhn and Feyerabend, that reasoning can be an effective motor of scientific change. There are limits to argument, but where those limits lie and what determines them is as yet an unresolved issue.

6.6 Galileo: From Aristotelianism to Circular Inertia

My aim in this section is to present a somewhat simplified account of the several stages in the development of Galileo's thought leading up to the publication, in the twilight of his life, of what appears to be a principle of rectilinear inertia. The seeds of the idea were sown in his twenties. Yet even though he spent much of his life steadily dismantling the Aristotelian system, Galileo did not succeed in freeing himself entirely from it. In fact I shall argue that it was precisely Aristotelian residues in his thought that inhibited his regarding rectilinear inertia as more than an abstract possibility. The discontinuous transition to inertial physics was a gradual and demanding process; as Clavelin has noted, perhaps somewhat harshly

> though Galileo succeeded in drawing conclusions that classical dynamics would fully endorse, he did so by means of a conceptual system that in no way foreshadowed that of classical dynamics ... Galileo could do no better than to erect a scaffolding, one that was bound to be removed just as soon as the new edifice had been completed.[48]

Galileo was born in 1564, and entered the University of Pisa in 1581. He was taught there both by orthodox Aristotelians and by men like Buonamico, who familiarized him with medieval impetus theory. Some of Galileo's very earliest work was purely Aristotelian and involved frequent allusions to the contributions of the fourteenth century Schoolmen at Oxford and at Paris. However, in about 1590, he wrote a treatize on motion (*De Motu*) in which he courageously criticized Aristotle. Drake is not alone in arguing that in this tract Galileo took the first essential step towards the concept of inertia.[49]

One of the most fundamental dichotomies in the Aristotelian system of thought was that between natural and violent motions. It was this that Galileo rejected in *De Motu*. On the one hand, he argued that these kinds of motions were not mutually exclusive: that there were motions which could be a mixture of both. On the other hand, and far more importantly, he argued that they were not mutually exhaustive either;[50] that there were motions which were neither natural nor violent. In a footnote he referred to these as 'neutral' motions.[51]

Being neither natural nor violent, neutral motions could persist of their own accord without the assistance of an inner or outer motor of any kind. Galileo suggested that a sphere rotating at the centre of the universe would move neutrally. So too would a round heavy body moving on a 'horizontal' plane — but only in the absence of external resistances (for example, with perfectly smooth and hard bodies and planes) and if the 'plane' was a *spherical surface*. On a real horizontal plane, on a plane tangential to the Earth's surface, the body moves upwards, away from the centre of the universe. To trigger its motion more than an 'arbitrarily minimal force'[52] was needed and, correlatively, its movement would not be neutral. Neutral motion was only possible on a surface each point of which was equidistant from the centre of the Earth (i.e. on a spherical surface, not on a horizontal plane).

These ideas were developed further by Galileo some years later in a treatize on mechanics, *Le Mecaniche*. This work dates from about 1595; it was revised around 1600. In it Galileo considered, inter alia, the downward movement of heavy bodies.[53] Traditionally, writers on dynamics had attributed this to a force inherent in the body itself — to its gravity. Gravity had two functions: it was identified with the *weight* of the body (i.e., the pressure exerted by the body when it was at rest) and it was also taken to be a *motor force* (i.e., it was the cause of the body's spontaneous downward movement to the centre of the Earth). Aristotle took the weight which impelled the body downwards to be its absolute weight. In the sixteenth century it was argued that the speed of the fall varied rather with the body's specific weight (i.e., with the quantity of matter contained in a given volume of the body). But all the same, weight still served both a gravific function and a motor function.

Now, analyzing the movement of a body down an inclined plain in *Le Mecaniche*, Galileo came to separate the motor function of gravity from gravity *qua* weight. He noted that as the angle of inclination of a plane diminished so too did the force required to keep a body in equilibrium on it. But of course the specific weight of the body did not change with the tilt of the plane. Galileo concluded that a body's weight was not a direct measure of the motor force impelling it downwards; that was measured by what he called a 'moment of descent'.

Consider now a heavy body moving down a plane that is initially vertical. As the angle of inclination decreases so too will the moment of descent — the force propelling the body down the plane. Finally, when the 'inclined' plane is horizontal, the moment of descent disappears: the body remains suspended, as it were, between motion and rest. Or as Galileo puts it

> On an exactly balanced surface the ball would remain indifferent and questioning between motion and rest, so that any the least force would be sufficient to move it, just as on the other hand any little resistance, such as that merely of the air that surrounds it, would be capable of holding it still. From this we may take the following conclusion as an indubitable axiom: That heavy bodies, all external and adventitious impediments being removed, can be moved in the plane of the horizon by any minimum force.[54]

What Galileo has done here is to reject the traditional idea that all matter has an innate tendency to come to rest. It will be remembered that for Aristotle to be heavy *meant* to move rectilinearly downwards; the natural place of heavy bodies was at the centre of the universe and there they would stay unless violently disturbed. Galileo, by distinguishing the motor function of gravity from gravity *qua* weight indicated that, at least in one situation, a heavy body did not necessarily tend to a state of rest, but could be 'indifferent and questioning between motion and rest'. Although this is not equivalent to regarding motion and rest as states in which a body persists of its own accord, it does locate them on the same ontological footing and as such is a substantial move away from Aristotelian dynamics.

The joint position which emerges from *De Motu* and *Le*

Mecaniche is that, in the absence of all forces, a heavy body will remain indifferent to motion and rest on a 'horizontal' surface, each point of which is equidistant from the centre of the Earth, and that its motion on such a surface will be 'neutral', requiring neither an intrinsic nor an extrinsic motor to sustain it. This is in fact the position which Galileo effectively adopts in one of his letters on sunspots written to Mark Welser in August 1612. After drawing an Aristotelian-sounding distinction between the natural and violent movement of heavy bodies, Galileo adds that

> Finally, to some movements they are indifferent, as are these same heavy bodies to horizontal motion, to which they have neither inclination (since it is not toward the center of the earth) nor repugnance (since it does not carry them away from that center). And therefore, all external impediments removed, a heavy body on a spherical surface concentric with the earth will be indifferent to rest and to movements toward any part of the horizon. And it will maintain itself in that state in which it has once been placed; that is, if placed in a state of rest, it will conserve it; and if placed in movement toward the west (for example), it will maintain itself in that movement. Thus a ship, for instance, having once received some impetus through the tranquil sea, would move continually around our globe without ever stopping; and placed at rest it would perpetually remain at rest, if in the first case all extrinsic impediments could be removed, and in the second case no external cause of motion were added.[55]

It must be stressed that Galileo's impetus is *not* that of Buridan. I quoted Buridan earlier (Section 6.2) as saying that, God having impressed an impetus on the celestial spheres when he created the world, 'these impetuses which he impressed in the celestial bodies were not decreased nor corrupted afterwards, because there was no inclination of the celestial bodies for other movements. Nor was there resistance which would be corruptive or repressive of that impetus'. Accordingly the celestial bodies moved perpetually in circles, *maintained by the impetus*. For Galileo a heavy body once set in motion under analogous circumstances *maintains itself* in movement. Buridan's motion is the effect of a 'permanent', (i.e., not self-expending) impetus which causes it. Galileo's motion is a state. He borrows the *word* but not the *concept* of impetus from his medieval predecessors.

The view that circular motion could be conserved by heavy bodies played an important part in Galileo's defence of

Copernicanism. In the *Dialogues Concerning the Two Chief World Systems* it was supplemented by the further assumption that circular motion was also the *natural* motion for the Earth and earthy objects: the diurnal rotation of the terrestrial globe was its 'own and natural motion, and hence that of all its parts, as a thing indelibly impressed upon them by nature'.[56] Terrestrial objects, in other words, have an innate tendency to rotate with the Earth about its centre in twenty-four hours and, in addition, to conserve any motion imparted 'violently' to them.

Attributing (natural) circular motion to the earthy was at once an assault on Aristotelianism, a crucial component in the development of Copernicanism, and a residue from traditional thought which impeded the development of a new science of motion. In the first place, it undermined the Aristotelian distinction between the celestial and the terrestrial. In Section 5.2 I showed how this distinction was based on a more fundamental dichotomy between circular and rectilinear motions, the only two kinds of motion which Aristotle took to be simple and natural. This, said Galileo, is the 'cornerstone, basis, and foundation of the entire structure of the Aristotelian universe, upon which are superimposed all other celestial properties'.[57] Yet, he insisted, it was defective: 'sounder architechtural precepts are required'. Galileo then went on to argue that 'only circular motion can naturally suit bodies which are integral parts of the universe as constituted in the best arrangement'.[58] The most that could be said for straight motion was that it restored order; it had no part whatsoever to play in sustaining it once it was established. On the contrary, once bodies were arranged in an orderly way, 'it is impossible that there should remain in them natural inclinations to move any more in straight motions, from which nothing would now follow but their removal from their proper and natural places; which is to say, their disordering'.[59] This was true both for celestial and terrestrial bodies: in a well ordered universe only circular motion could be natural and Aristotle had been inconsistent in arguing otherwise.

This was of course not the only way in which Galileo undermined the Aristotelian dichotomy between the heavens and the Earth. His telescopic observations led him to believe that the surface of the Moon, rather than being smooth and precisely spherical, as demanded by the Peripatetics, was 'not unlike the face of the Earth, relieved by chains of mountains and deep valleys';[60] that there were variable spots and blemishes on the face of the Sun, formerly believed to be ingenerable and incorruptible; and that the planets, distinguished by traditional cosmology as being 'luminous and resplendent' in contrast to the Earth, which was 'obscure and without light',[61] in fact had no luminosity of their own either. Yet it was the argument from cosmological order, critically appropriated from the Aristotelian heritage, it was his insistence that the terrestrial, too, moved naturally in circles, that helped Galileo to defuse Peripatetic objections to the Earth's diurnal motion.

Consider, for example, the tower argument. Sharing as it does in the natural movement of the Earth, the rock held at the top of a tower 'has as its primary tendency a revolution about the center of the whole in twenty-four hours, and it eternally exercises this natural propensity no matter where it is placed'.[62] We, too, share in the Earth's rotational movement though not, apparently, because our bodies partake of earthy matter, but because we are inhabitants of it and thus participate in its motion. Accordingly, 'whatever motion comes to be attributed to the earth must necessarily remain imperceptible to us and as if nonexistent, so long as we look only at terrestrial objects'.[63] Thus because the Earth shares its diurnal motion with the rock and ourselves, that motion will remain imperceptible to us and without any effect whatever. 'All that remains observable is the motion which we lack, and that is the grazing drop to the base of the tower'.[64]

The idea expressed in Galileo's relativity principle — that a shared motion is 'as if nonexistent' — involves a profound step beyond Aristotelian dynamics. For the Aristotelians motion was transitory, and expressed the nature of the moving body; accordingly various motions were only compatible with one another if they were also compatible with its nature. Galileo no longer makes this demand. He prises motion free from the body which experiences it, makes it an independent object of thought, and also makes it possible to compound motions geometrically. However, for all its novelty his innovation still remained anchored in tradition in that it presupposed a particular conception of cosmic order and perfection.

Galileo's attempted explanation of tidal movement as arising from a combination of the Earth's annual and diurnal movements also illustrates how, even while forging a radically new science of motion he fell back on traditional ideas. The arguments were deployed on the Fourth Day of the *Dialogue*, and were of singular importance in Galileo's eyes. His assaults on the Peripatetic objections to the Earth's movement made during the previous two days were not meant to establish Copernicanism. Galileo's aim was more modest: it was to show that terrestrial events which were ordinarily held to confirm the immobility of the Earth could as well be accommodated within the heliostatic system. For this reason he regarded his argument from the tides as 'the end and goal of all our previous discussions'.[65] It has been severely criticized,[66] in part because it apparently violates Galileo's own relativity principle. Brown, however, has argued that this is not so; that Galileo attributed different dynamical properties to Earth, Water and Air, and that Water, in particular, had no natural motion of its own (like the earthy) but could conserve an impressed circular motion. Galileo, he says, retains 'the core idea of the existence of different elements with different dynamical properties, but this idea is now absorbed into a new physics in which the two realms and the natural places of Aristotelian physics have been abandoned and new dynamical properties play the central role'.[67] That granted, Brown goes on to argue that the apparent inconsistency between Galileo's tidal theory and his theory of motion developed earlier in the *Dialogue* can be dissolved.

Brown notes, correctly, that there is some dispute among historians as to whether Galileo thought that a terrestrial body, once set in motion, would continue to move rectilinearly or circularly. In other words, it is not clear whether Galileo adopted a rectilinear or a circular principle of inertia. The source of the difficulty is usually located in the debate on motion in Galileo's *Dialogues Concerning Two New Sciences*, published in 1638. Discussing the movement of bodies on inclined planes, Galileo pointed out that

Furthermore we may remark that any velocity once imparted to a moving body will be rigidly maintained as long as the external causes of acceleration or retardation are removed, a condition which is found only on horizontal planes; for in the case of planes which slope downwards there is already present a cause of acceleration, while on planes sloping upward there is retardation; from this it follows that motion along a horizontal plane is perpetual; for, if the velocity be uniform, it cannot be diminished or slackened, much less destroyed.[68]

Galileo, then, asserts that a body once set in motion on a horizontal plane will move perpetually with a uniform velocity. This sounds very much as though he regarded inertial motion to be rectilinear. However, the assertion was qualified later in the *Dialogues* when Galileo discussed projectile motion: the motion of a body moving on a horizontal plane will only be uniform 'in the abstract'.[69] Since all bodies have an inherent tendency to move towards the centre of the Earth, uniform rectilinear motion is, strictly speaking, impossible. For as they move away from the point of contact of a horizontal plane tangential to the Earth' surface, their natural inclination to return to the centre will retard their motion. On the other hand, since in practice real horizontal planes are far shorter than the Earth's circumference, one can to all intents and purposes regard motion along them to be uniform, not retarded. Thus for the Galileo of the *Dialogues Concerning Two New Sciences*, inertial uniform motion becomes indistinguishable from rectilinear motion. However, given his lifelong commitment to the traditional view of gravity as an inherent property of bodies, this was only an abstract possibility, lacking adequate theoretical foundations. These foundations were provided by Newton.

Enmeshed as he was within the Aristotelian tradition, the horizons of Galileo's thought were limited by his past. As an agent immersed in the flux of history, his break with his heritage was neither clean nor complete. To have established a Newtonian science of motion free from cosmological premises, Galileo needed to locate his analysis of circular motion in a conceptual scheme which distinguished 'mass' from 'weight' and which had a correspondingly appropriate conception of gravitational attraction. This he did not do. As far as he was concerned gravity was an inherent property of heavy bodies, tending to impel them towards the centre of the Earth. No one had any idea what the 'essence' of gravity was; it was just a name, a household word.[70] Lacking an adequate theoretical understanding of gravity, Galileo could not abstract from it; his thought remained trapped at the level of our everyday experience of heavy bodies. And so, just as he came within a hairsbreadth of explicitly stating the principle of rectilinear inertia, he drew back, weighed down, as Marx put it, by 'the tradition of dead ages'.

That granted, it must also be stressed that Aristotelianism was not simply an epistemological obstacle to the development of inertial physics. In Chapter 1 I quoted Braudel to the effect that habits which imprison us and make decisions for us *also* help us to live. The Peripatetic system performed this double function for Galileo. On the one hand, it ultimately served to block his access to the principle of linear inertia. On the other hand, it provided him with invaluable resources on which he drew as he struggled to lay the foundations for a new science of motion.

There is little doubt, then, that Galileo's work was transitional, perhaps only the 'scaffolding' (Clavelin) erected while the foundations of classical physics were being laid, yet essential to the construction of the new edifice. It took time for Galileo to fight his way out of the maze of Aristotelian thought; even then, despite his immense efforts, he never managed to disentangle himself completely from it. But this does not mean, as is usually supposed, that the transition in question was continuous and evolutionary. On the contrary, it was part of a revolutionary transformation in which one system of thought ultimately replaced another *toto caelo*. It was discontinuous, but not sudden; it was revolutionary, but not necessarily unintelligible and arbitrary. Galileo attempted to argue his way out of Aristotelianism. He was not converted to a different view (Kuhn), nor did he try to convince his opponents by producing nonsense (Feyerabend).

The bankruptcy of these 'theories' of discontinuous transition reflects the subterranean pressures exerted by a crude and rigid distinction between, loosely, reason and emotion; the former calm, level-headed, continuous; the latter, at least during revolutionary periods, sudden, explosive, discontinuous. It may be thought that in stressing the importance of argument in Galileo's rejection of Aristotelianism I am, in fact, buttressing this dichotomy; that I am suggesting that the revolutionary change was rationally intelligible in the sense that reasoning or argument alone was sufficient to bring it about. But this is not so. Galileo argued for his position, of course. Whether or not others accepted his arguments is a different matter. The resistance that he encountered was immense, and Galileo was under no illusions as to the extent of the task with which he was confronted. After all his aim was to get his opponents to reject a system of thought which had survived for almost 2000 years. Sagredo, the educated layman in the *Dialogue Concerning the Two Chief World Systems* appreciated the risks that this involved. He spelt out Simplicio's fears, only to be reassured by Salviati

SAGR. I can put myself in Simplicio's place and see that he is deeply moved by the overwhelming force of these conclusive arguments. But seeing on the other hand the great authority that Aristotle has gained universally; considering the number of interpreters who have toiled to explain his meanings; and observing that the other sciences, so useful and necessary to mankind, base a large part of their value and reputation on Aristotle's credit; Simplicio is confused and perplexed, and I seem to hear him say, 'Who would there be to settle our controversies if Aristotle were to be deposed ? What other author should we follow in the schools, the academies, the universities ? . . . I pity him no less than I should some fine gentleman who, having built a magnificent palace at great trouble and expense, employing hundreds and hundreds of artisans, and then beholding it threatened with ruin because of poor foundations, should attempt, in order to avoid the grief of seeing the walls destroyed, adorned as they are with so many lovely murals; or the columns fall, which sustain the superb galleries, or the gilded beams; or the doors spoiled, or the pediments and the marble cornices, brought in at so much cost — should attempt, I say, to prevent the collapse with chains, props, iron bars, buttresses, and shores.

SALV. Well, Simplicio need not yet fear any such collapse; I undertake to insure him against damage at a much smaller cost. There is no danger that such a multitude of great, subtle, and wise philosophers will allow themselves to be overcome by one or two who bluster a bit. Rather, without even directing their pens against them, by means of silence alone, they place them in universal scorn and derision. It is vanity to imagine that one can introduce a new philosophy by refuting this or that author. It is necessary first to teach the reform of the human mind and to render it capable of distinguishing truth from falsehood, which only God can do. [71]

Galileo's pessimism regarding the limits of argument as a means of getting people to reject a view into which they have been socialized is not unfounded. In Chapter 4 I quoted Feyerabend as saying that 'We need to be virtually bludgeoned into detachment from our daily lives, our habits and mental laziness, which conceal from us the strangeness of the world' (see Chapter 4, Ref. 78). The resistance to change which makes this necessary is quite possibly unavoidable. Marris for one has pointed out that 'the impulse to defend the predictability of life is a fundamental and universal principle of human psychology'. All change, he notes, involves loss, and the sense of loss, or of impending loss, precipitates 'anxieties of change centred upon the struggle to defend or recover a meaningful pattern of relationships'. As human beings we have a 'deep-rooted and insistent need for continuity'.[72] This need emerges only too clearly in Galileo's attempt to disentangle himself from the structure of Peripatetic thought. It has also been noted by Marx, whom I quoted in Chapter 1 as saying that, just when people seem engaged in revolutionizing themselves and things, they 'anxiously conjure up the spirits of the past to their service', thereby creating the illusion that nothing has really changed.

If I am correct, it follows that as long as we identify reason with reasoning there is no hope of our theorizing discontinuous transitions as rationally intelligible. That granted, we must, I think, either become Dadaists or develop a conception of rationality which recognizes the limits of argument as a means of getting people to change their minds, and which takes cognisance of the material conditions under which argument is an effective motor of progress.

Notes

1 Galileo Galilei, in H. Crew and A. de Salvio (trans), *Dialogues Concerning Two New Sciences*, (Dover Publications, New York, 1954), pp.153–4.
2 Galileo Galilei, in S. Drake (trans), *Dialogue Concerning the Two Chief World Systems*, (University of California Press, Berkeley, second edition 1967), p.222.
3 J.H. Randall, 'The development of scientific method in the School of Padua', *Journal of the History of Ideas*, 1 (1940), 178–9.
4 P. Duhem, quoted in M. Clavelin, *The Natural Philosophy of Galileo*, A.J. Pomerans (trans), (MIT Press, Cambridge, Mass. and London, 1974), p.94.
5 A. Maier, quoted in S. Drake, *Galileo Studies* (University of Michigan Press, Ann Arbor, 1970), p.246.

6 *Ibid.* p.246.
7 *Ibid.* p.20.
8 A. Koyré, *Galileo Studies*, J. Mepham (trans), (Harvester Press, Hassocks, 1978), p.3.
9 A. Koyré, *Metaphysics and Measurement* (Chapman & Hall, London, 1968), pp.21–2.
10 Aristotle, quoted in M. Clagett, *The Science of Mechanics in the Middle Ages* (Oxford University Press, London, 1959), p.506.
11 J. Buridan, quoted in M. Clagett, *op.cit.* pp.534–5.
12 *Ibid.* p.535.
13 *Ibid.* p.536.
14 *Ibid.* p.536.
15 A.R. Hall, *The Scientific Revolution, 1500–1800* (Longmans, Harlow, 1954), pp.247–8.
16 *Ibid.* p.248.
17 A. Koyré, *Newtonian Studies* (University of Chicago Press, Chicago, 1965), p.15.
18 Toulmin regards the Aristotelian theory of natural motion to be an example of what he calls an 'ideal of natural order'. He has argued that one of the central aims of science is to render the course of nature not merely predictable, but intelligible, 'and this has meant looking for rational patterns of connections in terms of which we can make sense of the flux of events' (see S. Toulmin, *Foresight and Understanding* (Hutchinson, London, 1961), p.99). Ideals of natural order serve this purpose. They express those regularities which people come to take for granted and not in need of further explanation. Toulmin stresses that although people come to take them for granted they were not always adopted unquestioningly. We are justified in adhering to them and in 'placing the trust in them that we do, only because — and to the extent that — they have proved their worth in competition with alternatives' (S. Toulmin, *op.cit.* p.102).
19 This is a translation from the original Latin as made by A. Motte in 1729. See A. Koyré, *Newtonian Studies*, *op.cit.* p.66.
20 The discussion which follows draws heavily on Koyré's analyses. See, for example, his *Newtonian Studies*, Chapters 1 and 3, and his *Metaphysics*, Chapters 1 and 2.
21 A. Koyré, *Galileo Studies*, *op.cit.* p.130. See also A. Koyré, *Metaphysics*, *op.cit.* p.4.
22 M. Clagett, p.523. In this paragraph and the next the views I am attributing to Clagett were anticipated by Duhem.
23 *Ibid.* p.539.
24 *Ibid.* p.540.
25 D. Shapere, 'Meaning and scientific change', in R. Colodny (ed), *Mind and Cosmos: Explorations in the Philosophy of Science* (University of Pittsburgh Press, Pittsburgh, 1966), p.72.
26 *Ibid.* p.79.
27 D. Shapere, *Galileo* (University of Chicago Press, Chicago, 1974), pp.52–3. It is noteworthy that Shapere omits to mention the idea that gravity was an inherent property of (heavy) bodies as being an obstacle

to the development of the principle of (rectilinear) inertia. For further details, see Section 6.6.
28   M. Clavelin, op.cit. p.47.
29   D. Shapere, Galileo, op.cit. p.89.
30   Ibid. p.109. Original emphasis suppressed.
31   D. Shapere, in R. Colodny (ed), op.cit. p.79.
32   J. Dunn, 'The identity of the history of ideas', Philosophy, 43 (1968), 92.
33   Ibid. p.99. My emphasis.
34   W. Dray, Laws and Explanation in History (Oxford University Press, London, 1957), Chapter 5.
35   J. Dunn, op.cit. p.96.
36   S. Drake, 'Impetus theory reappraised', Journal of the History of Ideas, 36 (1975), 27.
37   K.R. Popper, Objective Knowledge (Clarendon Press, Oxford, 1972), Chapter 3.
38   D. Shapere, in R. Colodny (ed), op.cit. p.76. By and large I am not convinced by most of the examples which Shapere brings in support of this claim. The claim itself seems to me to be important, though.
39   M. Clavelin, op.cit. p.103.
40   S. Drake, Galileo Studies, op.cit. p.244.
41   P.K. Feyerabend, 'Changing patterns of reconstruction', British Journal for the Philosophy of Science, 28 (1977), 363.
42   T.S. Kuhn, The Structure of Scientific Revolutions (University of Chicago Press, Chicago, second edition, 1970), pp.97–8.
43   P.K. Feyerabend, op.cit. pp.364, 365.
44   P.K. Feyerabend, Against Method (New Left Books, London, 1975), p.285.
45   T.S. Kuhn, op.cit. p.150.
46   P.K. Feyerabend, Against Method, op.cit. p.270.
47   D. Shapere, in R. Colodny (ed), op.cit. pp.68, 79.
48   M. Clavelin, op.cit. p.357.
49   S. Drake, Galileo Studies, op.cit. p.249.
50   Ryle has pointed out how personal change is inhibited when individuals conceptualize their situation in terms of false dichotomies which restrict their range of behaviour. This leads to a loss of sense of agency and self-efficacy. An example which he gives is 'In relationships I am *either* close to someone and feel smothered, *or* I am cut off and feel lonely' (See A. Ryle, 'The focus in brief interpretative psychotherapy: dilemmas, traps and snags as target problems', British Journal of Psychiatry, forthcoming). I suspect that there is an important link between change in general and the suspension of alternatives previously felt to be mutually exclusive or mutually exhaustive. It is perhaps primarily by superseding *them* that those spaces are created in which growth is possible.
51   Galileo Galilei, On Motion, in I.E. Drabkin and S. Drake (trans), On Motion and On Mechanics (University of Wisconsin Press, Madison, 1960), p.67.
52   Ibid. p.68.
53   My treatment here follows that of M. Clavelin, op.cit. pp.165–74.

54 Galileo Galilei, *On Mechanics*, in I.E. Drabkin and S. Drake (trans), *op.cit*. p.171.
55 Galileo Galilei, *Letters on Sunspots*, in S. Drake (trans), *Discoveries and Opinions of Galileo* (Doubleday Anchor, New York, 1957), pp.113–4.
56 Galileo Galilei, in S. Drake (trans), *Dialogue Concerning the Two Chief World Systems*, *op.cit*. p.142.
57 *Ibid*. p.18.
58 *Ibid*. p.32.
59 *Ibid*. p.20.
60 Galileo Galilei, *The Starry Messenger*, in S. Drake (trans), *Discoveries and Opinions of Galileo*, *op.cit*. p.31.
61 M. Clavelin, *op.cit*. p.199.
62 Galileo Galilei, in S. Drake (trans), *Dialogue Concerning the Two Chief World Systems*, *op.cit*, p.142.
63 *Ibid*. p.114.
64 *Ibid*. p.171.
65 *Ibid*. p.415.
66 See, for example, W.R.J. Shea, 'Galileo's claim to fame: the proof that the Earth moves from the evidence of the tides', *British Journal for the History of Science*, 5 (1970), 111–27, and the references therein.
67 H.I. Brown, 'Galileo, the elements, and the tides', *Studies in the History and Philosophy of Science*, 7 (1976), 349.
68 Galileo Galilei, in H. Crew and A. de Salvio (trans), *op.cit*. p.215.
69 *Ibid*. p.251.
70 Galileo Galilei, in S. Drake (trans), *Dialogue Concerning the Two Chief World Systems*, *op.cit*. p.234.
71 *Ibid*. pp.56–7.
72 P. Marris, *Loss and Change* (Routledge & Kegan Paul, London, 1978), pp. 1, 2.

7 CONCLUDING REMARKS

In a nutshell, my aim in this study has been to defend, perhaps even to rehabilitate, the epistemology of rejection and refusal. I have argued that the view that growth and progress can be achieved steadily and cumulatively within a more or less stable and unquestioned framework or pattern of relationships is at best only a half truth. There are times when the framework becomes a constricting cage, when all the possibilites of development within it have been exhausted. At such times it is *impossible* to move forward without *dismantling* the framework, opposing and rejecting it, and replacing it with another.

It is to Popper's lasting credit that he saw growth as requiring opposition, and that he attempted to develop a theory of rational criticism. It is also to his credit that, having a sufficiently astute sense of history, he realized that an entire established order cannot be rejected in one fell swoop, that we cannot start from scratch. As social beings, historical agents are themselves immersed in history whether they like it or not, and whether they realize it or not. During socialization the past seeps into them; it lives on in them, makes decisions for them, shapes their habits, and limits their horizons. To start from scratch, to begin the race where Adam and Eve began, as Popper puts it, is not simply to wipe out a prevailing structure of relationships: it is also to destroy oneself.

The value of these insights is regrettably dissipated by the presuppositional framework within which Popper theorizes the growth of knowledge. Informed by a rigid distinction between psychology and logic, and between discovery and justification, he locates the motor of development in the arbitrary and the non-rational. Flashes of insight, acts of poetic intuition: it is from these sources that new ideas arise. Disengaging themselves from the present, soaring above and beyond it, it is creative and imaginative individuals who provide the raw material for further growth. That material is fed into the scientific community, where it is processed, criticized, and is either rejected as false and eliminated or woven into the existing fabric of thought. Revolutions shrink to exhilarating moments of individual consciousness; all real change is continuous. The creative individual is another craftsman adding a buttress here, a gargoyle there to the cathedral of science, the foundations of which remain firm come what may.

Lakatos and, more overtly, Feyerabend, resent this. They resent the restrictions on individual liberty which Popper's falsificationism and his demand for theory elimination imply. Together they have conclusively exposed the weaknesses of his position. Lakatos emphasized that theories have a history, and that to eliminate them too rapidly would be to destroy them prematurely. He then landed up in the embarassing position of not knowing quite how to eliminate them at all. As a last resort he rather despairingly appealed to the not so subtle pressure that could be applied by the members of the scientific establishment. It was on their shoulders that he placed the burden of clearing the historical stage of a defunct and outworn system of thought — which may have recovered and blossomed if only . . . .

Feyerabend will have none of this. His central value is individual happiness, by which he means that everyone should be allowed to follow their own inclinations. Like Popper and Lakatos he sees the struggle for autonomy as one in which the individual tries to step outside and rise above the stifling present: new ideas are drawn from the realms of fantasy, of the imagination, from repressed layers of consciousness, from the almost forgotten archives of the past. They are developed and elaborated in a space alongside prevailing orthodoxy, which is criticized only insofar as it aims to smother the novel. It is not rejected and dismantled; merely held at bay. Within such spaces grow a rich and diverse plurality of world views. Ideally any particular view coexists peacefully with the others, and will last until no one finds it attractive or pleasing any longer. Thereupon it will disappear, withering away more as a result of collective indifference than because of any basic flaw in it.

There is a steady progression from Popper to Feyerabend away from a critical, oppositional epistemology towards a conservative, relativistic position which tolerates everything, including the intolerable. Feyerabend is quite explicit about this: he has nothing but contempt for contemporary science but in his ideal world he wouldn't dream of stopping anyone from pursuing it if they liked. Along with the drift to relativism goes an assault on the notions of truth and falsehood, and an increasing emphasis on the merits of flippancy, even superficiality. Argument is devalued and reduced to the level of rationalization; anger is dismissed as symptomatic of puritanical fervour.

To appreciate the significance of Feyerabend's position it is important to locate it in its own historical context. For many years now Popper has attempted to lay down the conditions under which it is rational to reject a scientific theory. On investigating the actual practice of scientists, however, Kuhn discovered that they simply did not behave as Popper said they should or, more precisely, as he thought that Popper had said they should. If Popper was correct, Kuhn argued, most scientists didn't behave rationally at all, and in particular they didn't behave rationally during revolutionary periods — which were Popper's special concern. Lakatos was profoundly disturbed by this. He jettisoned a simple version of Popperian rationality, and located within Popper's work another strand which seemed able to save scientists from charges of irrational behaviour. Along with this he suggested that the history of science become a laboratory in which philosopher's theories of rationality be tested against actual scientific practice, and be revized if found wanting. Feyerabend took Lakatos at his word. Analyzing Galileo's behaviour he concluded that none of the existing theories of rationality could accommodate it; revolutions appeared once again to be fundamentally irrational and passionate outbursts, driven by the motor of personal gain.

A major theme of this study has been that to avoid Feyerabend's conclusions there is no point in tinkering around with the pre-suppositional framework which he shares (or at least does not systematically criticize) with Popper and with Lakatos. It is the framework itself that must be rejected, at least if one wants to develop a theory of rational scientific growth which takes cognisance of the actual behaviour of scientists as historical agents. The reason is clear: by rigidly demarcating the context of discovery from the context of justification, and restricting themselves to the latter,

CONCLUDING REMARKS 217

philosophers are forced to stand outside history, as spectators. At the same time, in attempting seriously to theorize the actual practice of scientists they are compelled to take a participant's perspective. Torn by this internal contradiction, their enterprize is fatally flawed, incoherent, and doomed to failure. This is not to say, of course, that the immense efforts of Popper, Lakatos and Feyerabend have been in vain. On the contrary, they have been invaluable. For they have revealed just how important it is to take a participant's perspective on the historical process if one wishes to develop a coherent conception of rational scientific growth, and of revolutionary change in particular.

Taking Galileo's perspective on the Scientific Revolution amounts to recognizing that, far from being benignly tolerant of Aristotelianism — the prevailing orthodoxy of the day — he was uncompromizingly hostile to it. He opposed it, and set out systematically to demolish it, believing that the Earth-centred approach to astronomy was 'undoubtedly false', and that it was 'impossible' to accommodate the Copernican hypothesis within the framework of Peripatetic thought. In Galileo's case this understanding arose, at least in part, from a theoretical analysis of the pattern of relationships constituting that structure, an analysis which revealed the complex interconnection between its several elements. The structure was rejected because Galileo, and perhaps Copernicus too, came to believe that the problem of planetary motion not only *had not*, but actually *could not*, be solved within the limits which it imposed. The growing awareness that it is impossible to advance without dismantling an existing edifice is itself an historical process, the precise determinants of which I have not investigated here.

Revolutionary changes are not effected by escaping from the confines of the present into the realm of the non-rational, into the realms of fantasy, of the imagination, of poetic intuition. It is not by abstracting oneself from a prevailing pattern of relationships, by soaring above and beyond them, that one transforms them. Only oppositional engagement can do that. But this is not to say that there is nothing to be gained from building up alternatives to a prevailing orthodoxy which survive alongside it. Such alternatives open new vistas, exposing briefly what would have otherwise remained unrecognized possibilities. But they have their limits. For only those possibilities can be actualized within them which do not require the transformation of the dominant pattern of relationships. That pattern remains as an ever-present constraint on what can be achieved in pockets alongside it. In turn then its limits are exposed and gradually the impossibility of continuing within its strictures becomes manifest.

Revolutionary transformations of the human understanding are neither cataclysmic nor do they demolish an existing structure in one blinding flash. They develop unevenly. Having decided that it is impossible to make any further progress within the confines of an existing system, the agents of revolutionary change sow the seeds of a new order in the harsh ground of the old. Initially the losses are tremendous. The stability and the security derived from moving within the limits of a well established, deeply entrenched, and richly articulated system are sacrificed. It is only the conviction that it is not possible to move forward without dismantling that system, together with the promise of what a new future holds, which sustains people during such unsafe and disorienting periods. Any attempt to theorize discontinuous change as rational must recognize this double aspect of revolutionary commitment.

The optimism of the agents of revolutionary change may be necessary if they are to do what must be done. Yet they too are immersed in history, and are constrained by the past even as they rebel against it. The pattern of relationships which they set out to dismantle is both a framework from which they strive to disentangle their new project and a snare which unwittingly subverts their efforts to do so. As such it shackles the realization of some of the possibilities inherent in the new blossoming system. These remain purely abstract (as did the principle of linear inertia in Galileo's work) unless and until the appropriate material conditions prevail. In the particular case of the science of mechanics, this meant purging dynamics of Aristotelian residues, and developing the concepts of 'mass' and 'force'.

On the other hand, Aristotelianism was not simply an epistemological obstacle which subverted the development of inertial physics. It also provided Galileo with notions like 'impetus' which played an essential and productive role in the articulation of the new science of motion. Thus, even as he rebelled against Peripatetic thought Galileo extracted elements from it which helped him to construct his alternative.

Continuist history recognizes that historical agents are constrained by the past. Thereupon it searches eagerly for anticipations, analogues and resemblances, levelling out differences and papering over cracks. It fails to appreciate that revolutions take time to consolidate themselves, and that the struggle to lay down a new network of viable relationships is a long, painful and protracted one. Trapped at the level of appearances, it turns a blind eye to the destructive forces which are smouldering beneath the surface of thought and of action, and which periodically disturb its tranquility. Discontinuist history is sensitive to differences, seeks to explain them, and to identify the deep structures of which they are the manifest effects. It is patient. It realizes that decades, perhaps centuries, are needed for one universe of discourse or pattern of relationships to replace a rival.

The conception of revolution that I have defended here takes it that such changes are discontinuous in that they involve the rejection and replacement of an established system of thought or pattern of relationships by a new and better alternative. However, against the 'liberal' conception of revolution I have stressed that rejection and replacement do not occur simultaneously. With Feyerabend, I have suggested that science develops unevenly. The agents of revolutionary change reject a fully-fledged system and espouse their alternative on the basis of future promise as much as, if not more than, on present success. Both Kuhn and Feyerabend recognize this latter, liberating yet unsettling dimension of revolutionary change. Thus Kuhn writes

The man who embraces a new paradigm at an early stage must often do so in defiance of the evidence provided by problem-solving. He must, that is, have faith that the new paradigm will succeed with the many large problems that confront it, knowing only that the older paradigm has failed with a few. A decision of that kind can only be made on faith.[1]

Where Kuhn speaks of faith, Feyerabend speaks of inclination and interest. Perhaps the single most important way in which I differ from them lies in my refusal, at this stage, to concede that much to the non-rational. The confidence which people like Copernicus and Galileo needed to defend the heliostatic system was derived, at least in part, from their rejection of Aristotelianism. To my mind, it is essential that that be recognized and that it be built into an assessment of the rationality of their commitment, and of revolutionary commitment in general.

Notes

1   T.S. Kuhn, *The Structure of Scientific Revolutions* (University of Chicago Press, Chicago, second edition, 1970), p.158.

BIBLIOGRAPHY

L. Althusser, *For Marx*, B. Brewster (trans), (Penguin, Harmondsworth, 1969).
L. Althusser and E. Balibar, *Reading Capital*, B. Brewster (trans), (New Left Books, London, 1970).
S.A. Bedini, 'The instruments of Galileo Galilei', in E. McMullin (ed), *Galileo: Man of Science* (Basic Books, New York, 1967), pp.256–92.
R. Bhaskar, *A Realist Theory of Science* (Harvester Press, Hassocks, 1978).
R. Bhaskar, 'Feyerabend and Bachelard: two philosophers of science', *New Left Review*, No. 94 (1975), pp.31–55.
R. Blackburn, *Revolution and Class Struggle* (Fontana, London, 1977).
H.I. Brown, 'Galileo, the elements and the tides', *Studies in the History and Philosophy of Science*, 7 (1976), 337–51.
H. Butterfield, *The Whig Interpretation of History* (Bell, London, 1931).
H. Butterfield, *The Origins of Modern Science 1300–1800* (Bell, London, 1968).
E.H. Carr, *What is History* ? (Penguin, Harmondsworth, 1964).
A.F. Chalmers, *What is This Thing Called Science* ? (University of Queensland Press, St. Lucia, 1976).
M. Clagett, *The Science of Mechanics in the Middle Ages* (Oxford University Press, London, 1959).
M. Clavelin, *The Natural Philosophy of Galileo*, A.J. Pomerans (trans), (MIT Press, Cambridge, Mass., and London, 1974).
I.B. Cohen, 'The eighteenth-century origins of the concept of scientific revolution', *Journal of the History of Ideas*, 37 (1976), 257–88.
R.S. Cohen, P.K. Feyerabend and M.W. Wartofsky (eds), *Boston Studies in the Philosophy of Science*, Vol. XXXIX, Essays in Honour of Imre Lakatos (Reidel, Dordrecht, 1976).
A.C. Crombie, *Augustine to Galileo* (Heinemann, London, 1952).
J. Curthoys and W. Suchting 'Feyerabend's discourse against method: a Marxist critique', *Inquiry*, 20 (1977), 243–397.
E.J. Dijksterhuis, 'The origins of classical mechanics from Aristotle to Newton', in M. Clagett (ed), *Critical Problems in the History of Science* (University of Wisconsin Press, Madison, 1959), pp.163–84.
E.J. Dijksterhuis, *The Mechanization of the World Picture*, C. Dikshoorn (trans), (Oxford University Press, London, 1961).
S. Drake, *Discoveries and Opinions of Galileo*, (Doubleday Anchor Books, New York, 1957).
S. Drake, *Galileo Studies* (University of Michigan Press, Ann Arbor, 1970).
S. Drake, 'Mathematics and discovery in Galileo's physics', *Historia Mathematica*, 1 (1974), 129–50.
S. Drake, 'Galileo's new science of motion', in M.L. Righini Bonelli and W.R. Shea (eds), *Reason, Experiment, and Mysticism in the Scientific Revolution* (Science History Publications, New York, 1975), pp.131–56.

S. Drake, 'Impetus theory reappraised', *Journal of the History of Ideas*, 36 (1975), 27–46.

S. Drake, 'Free fall from Albert of Saxony to Honoré Fabri', *Studies in the History and Philosophy of Science*, 5 (1975), 347–66.

S. Drake, 'The evolution of *De Motu*', *Isis*, 67 (1976), 239–50.

S. Drake, 'A further reappraisal of impetus theory', *Studies in the History and Philosophy of Science*, 7 (1976), 319–36.

S. Drake and C.D. O'Malley, *The Controversy on the Comets of 1618* (University of Pennsylvania Press, Philadelphia, 1960).

W. Dray, *Laws and Explanation in History* (Oxford University Press, London, 1957).

P. Duhem, *The Aim and Structure of Physical Theory* (Princeton University Press, Princeton, 1954).

P. Duhem, *To Save the Phenomena* (University of Chicago Press, Chicago, 1969).

J. Dunn, 'The identity of the history of ideas', *Philosophy*, 43 (1968), 85–104.

B. Easlea, *Liberation and the Aims of Science* (Chatto & Windus/Sussex University Press, London, 1973).

R. Edgley, *Reason in Theory and Practice* (Hutchinson, London, 1969).

R. Edgley, 'Reason and violence', *Radical Philosophy*, No. 4 (Spring 1973), 18–24.

R. Edgley, 'The romantic energy', *Essays in Criticism*, XXV (1975), 261–71.

R. Edgley, 'Philosophy in Academia', *Oxford Review of Education*, 1 (1975), 203–10.

R. Edgley, 'Science, social science and socialist science: reason as dialectic', *Radical Philosophy*, No. 15 (Autumn 1976), 2–7.

R. Edgley, 'Dialectic: the contradiction of Colletti', *Critique*, No. 7 (Winter 1976/7), 47–52.

L.S. Feuer, *Einstein and the generations of science* (Basic Books, New York, 1974).

P.K. Feyerabend, 'Explanation, reduction and empiricism', in H. Feigl and G. Maxwell (eds), *Minnesota Studies in the Philosophy of Science*, Vol. III (University of Minnesota Press, Minneapolis, 1962), pp.28–97.

P.K. Feyerabend, 'How to be a good empiricist — A plea for tolerance in matters epistemological', in B. Baumrin (ed), *Philosophy of Science. The Delaware Seminar*, Vol. 2, 1962–1963 (Interscience, New York, 1963), pp.3–39.

P.K. Feyerabend, 'Realism and instrumentalism: comments on the logic of factual support', in M. Bunge (ed), *The Critical Approach to Science and Philosophy* (The Free Press, Glencoe, 1964), pp.280–308.

P.K. Feyerabend, 'Problems of microphysics', in R.G. Colodny (ed), *Frontiers of Science and Philosophy* (Allen & Unwin, London, 1964), pp.189–283.

P.K. Feyerabend, 'Reply to criticism', in R.S. Cohen and M.W. Wartofsky (eds), *Boston Studies in the Philosophy of Science*, Vol. 2 (Humanities Press, New York, 1965), p.223–61.

P.K. Feyerabend, 'Problems of empiricism', in R.G. Colodny (ed), *Beyond the Edge of Certainty* (Prentice-Hall, New Jersey, 1965), pp.145–260.

P.K. Feyerabend, 'On the improvement of the sciences and arts, and the possible identity of the two', in R.S. Cohen and M.W. Wartofsky (eds), *Boston Studies in the Philosophy of Science*, Vol. III (Reidel, Dordrecht, 1967), pp.387–415.

P.K. Feyerabend, 'The Theatre as an instrument of the criticism of ideologies', *Inquiry*, 10 (1967), 298–312.

P.K. Feyerabend, 'Outline of a pluralistic theory of knowledge and action', in S. Anderson (ed) *Planning for Diversity and Choice* (MIT Press, Cambridge, Mass., 1968), pp.275–84.

P.K. Feyerabend, 'Classical empiricism', in R.E. Butts and J.W. Davis (eds), *The Methodological Heritage of Newton* (Blackwells, Oxford, 1970), pp.150–70.

P.K. Feyerabend, 'Problems of empiricism, Part II', in R.G. Colodny (ed), *The Nature and Function of Scientific Theories* (University of Pittsburgh Press, Pittsburgh, 1970), pp.275–353.

P.K. Feyerabend, 'Consolations for the specialist', in I. Lakatos and A. Musgrave (eds), *Criticism and the Growth of Knowledge* (Cambridge University Press, Cambridge, 1970), pp.197–230.

P.K. Feyerabend, 'Against method: outline of an anarchistic theory of knowledge', in M. Radner and S. Winokur (eds), *Minnesota Studies in the Philosophy of Science*, Vol. IV (University of Minnesota Press, Minneapolis, 1970), pp.17–130.

P.K. Feyerabend, 'Philosophy of science: a subject with a great past', in R.H. Steuwer (ed), *Minnesota Studies in the Philosophy of Science*, Vol. V (University of Minnesota Press, Minneapolis, 1970), pp.172–83.

P.K. Feyerabend, 'Zahar on Einstein', *British Journal for the Philosophy of Science*, 25 (1974), 25–8.

P.K. Feyerabend, 'Popper's *Objective Knowledge*', *Inquiry*, 17 (1974), 475–507.

P.K. Feyerabend, *Against Method* (New Left Books, London, 1975).

P.K. Feyerabend, 'Let's make more movies', in C.J. Bontempo and S.J. Odell (eds), *The Owl of Minerva* (McGraw-Hill, New York, 1975), pp.201–10.

P.K. Feyerabend, 'Imre Lakatos', *British Journal for the Philosophy of Science*, 26 (1975), 1–18.

P.K. Feyerabend, 'Science: the myth and its role in society', *Inquiry*, 18 (1975), 167–181.

P.K. Feyerabend, 'How to defend society against science', *Radical Philosophy*, No. 11 (Summer 1975), 3–8.

P.K. Feyerabend, 'On the critique of scientific reason', in R.S. Cohen, P.K. Feyerabend, and M.W. Wartofsky (eds), *Boston Studies in the Philosophy of Science*, Vol. 39 (Reidel, Dordrecht, 1976), pp.109–43. Also in C. Howson (ed), *Method and Appraisal in the Physical Sciences* (Cambridge University Press, Cambridge, 1976), pp.309–39.

P.K. Feyerabend, 'Logic, literacy and Professor Gellner', *British Journal for the Philosophy of Science*, 27 (1976), 381–91.

P.K. Feyerabend, 'Changing patterns of reconstruction', *British Journal for the Philosophy of Science*, 28 (1977), 351–82.

P.K. Feyerabend, 'From incompetent professionalism to professionalized incompetence — the rise of a new breed of intellectuals', *Philosophy of the*

Social Sciences, 8 (1978), 37–53.
M. Foucault, *The Archaeology of Knowledge* (Tavistock Publications, London, 1974).
M. Foucault, *The Order of Things* (Tavistock Publications, London, 1974).
M. Foucault, 'Human nature: justice versus power', debate with N. Chomsky in F. Elders (ed), *Reflexive Waters* (Souvenir Press, London, 1974).
Galileo Galilei, *The Assayer*, in S. Drake and C.D. O'Malley (trans), *The Controversy on the Comets of 1618* (University of Pennsylvania Press, Philadelphia, 1960), pp.151–335.
Galileo Galilei, *The Assayer*, in S. Drake (trans), *Discoveries and Opinions of Galileo* (Doubleday Anchor, New York, 1957), pp.229–80.
Galileo Galilei, *Dialogue Concerning the Two Chief World Systems: Ptolemaic and Copernican*, S. Drake (trans), (University of California Press, Berkeley, second edition, 1967).
Galileo Galilei, *Dialogues Concerning Two New Sciences*, H. Crew and A. de Salvio (trans), (Dover Publications, New York, 1954).
Galileo Galilei, *Letters on Sunspots*, in S. Drake (trans), *Discoveries and Opinions of Galileo* (Doubleday Anchor, New York, 1957), pp.87–144.
Galileo Galilei, *On Mechanics*, in I.E. Drabkin and S. Drake (trans), *On Motion and on Mechanics* (University of Wisconsin Press, Madison, 1960), pp.147–86.
Galileo Galilei, *On Motion*, in I.E. Drabkin and S. Drake (trans), *On Motion and on Mechanics* (University of Wisconsin Press, Madison, 1960), pp.13–131.
Galileo Galilei, *The Starry Messenger*, in S. Drake (trans), *Discoveries and Opinions of Galileo* (Doubleday Anchor Books, New York, 1957), pp.21–58.
S.W. Gaukroger, 'Bachelard and the problem of epistemological analysis', *Studies in the History and Philosophy of Science*, 7 (1976), 189–244.
E. Gellner, 'Beyond truth and falsehood', *British Journal for the Philosophy of Science*, 26 (1975), 331–42.
A. Gerschenkron, 'On the concept of continuity in history', in his *Continuity in History and Other Essays* (Harvard University Press, Cambridge, Mass., 1968), pp.11–39.
J. Giedymin, 'The paradox of meaning variance', *British Journal for the Philosophy of Science*, 21 (1970), 257–68.
O. Gingerich, 'Dissertatio cum Professore Righini et Sidereo Nuncio', in M.L. Righini Bonelli and W.R. Shea (eds), *Reason, Experiment, and Mysticism in the Scientific Revolution* (Science History Publications, New York, 1975), pp.77–88.
A.R. Hall, *The Scientific Revolution, 1500–1800* (Longmans, Harlow, 1954).
A.R. Hall, 'Magic, metaphysics and mysticism in the scientific revolution', in M.L. Righini Bonelli and W.R. Shea (eds), *Reason, Experiment, and Mysticism in the Scientific Revolution* (Science History Publications, New York, 1975), pp.275–82.
S. Halper and J.R. Sterling (eds), *Latin America: The Dynamics of Social Change* (Allison & Busby, London, 1972).
N.R. Hanson, *Patterns of Discovery* (Cambridge University Press,

Cambridge, 1958).
L.O. Kattsoff, 'Ptolemy and scientific method', *Isis*, 38 (1947), 18–22.
R. Keat and J. Urry, *Social Theory as Science* (Routledge & Kegan Paul, London, 1975).
S. Körner, 'Mathematical frameworks in scientific thinking', *Advancement of Science* (March 1968), 306–10.
A. Koyré, *From the Closed World to the Infinite Universe* (John Hopkins Press, Baltimore, 1957).
A. Koyré, 'Influence of philosophic trends on the formulation of scientific theories', in P.G. Frank (ed), *The Validation of Scientific Theories* (Collier Books, New York, 1961).
A. Koyré, *Newtonian Studies* (University of Chicago Press, Chicago, 1965).
A. Koyré, *Metaphysics and Measurement* (Chapman & Hall, London, 1968).
A. Koyré, *The Astronomical Revolution* (Methuen, London, 1973).
A. Koyré, *Galileo Studies*, J. Mepham (trans), (Harvester Press, Hassocks, 1978).
J. Krige, 'Graham on the logic of belief', *Inquiry*, 19 (1976), 355–60.
J. Krige, 'A critique of Popper's conception of the relationship between logic, psychology and a critical epistemology', *Inquiry* 21 (1978), 313–35.
J. Krige, 'Popper's epistemology and the autonomy of science', *Social Studies of Science*, 8 (1978), 287–307.
J. Krige, 'What's so great about facts ?', in J. Irvine, I. Miles and J. Evans (eds), *Demystifying Social Statistics* (Pluto Press, London, 1979), pp.53–62.
S. Kruks, 'The philosophy of Merleau-Ponty', *Radical Philosophy*, No. 11 (Summer 1975), 17–24.
T.S. Kuhn, *The Copernican Revolution* (Harvard University Press, Cambridge, Mass., 1957).
T.S. Kuhn, *The Structure of Scientific Revolutions* (University of Chicago Press, Chicago, second edition, 1970).
T.S. Kuhn, 'Notes on Lakatos', in R.C. Buck and R.S. Cohen (eds), *Boston Studies in the Philosophy of Science*, Vol. VIII, 1970 (Reidel, Dordrecht, 1971), pp.137–46.
I. Lakatos, 'Changes in the problem of inductive logic', in I. Lakatos (ed), *The Problem of Inductive Logic* (North-Holland, Amsterdam, 1968), pp.315–417.
I. Lakatos, 'A letter to the director of the London School of Economics', dated 28th March 1968, in C.B. Cox and A.E. Dyson (eds), *Fight for Education: A Black Paper* (The Critical Quarterly Society, London), pp.27–31.
I. Lakatos, 'Criticism and the methodology of scientific research programmes, *Proceedings of the Aristotelian Scoeity*, 69 (1968–9), 149–86.
I. Lakatos, 'Falsification and the methodology of scientific research programmes', in I. Lakatos and A. Musgrave (eds), *Criticism and the Growth of Knowledge* (Cambridge University Press, Cambridge, 1970), pp.91–196.
I. Lakatos, 'History of science and its rational reconstructions', in R.C. Buck and R.S. Cohen (eds), *Boston Studies in the Philosophy of Science*, Vol. VIII, 1970 (Reidel, Dordrecht, 1971), pp.91–136; and 'Replies to critics', pp.174 *et seq*.

I. Lakatos, 'Science and pseudo-science', BBC Talk, Open University, 1973.
I. Lakatos, 'Popper on demarcation and induction', in P.A. Schilpp (ed), *The Philosophy of Karl Popper* (Open Court, La Salle, 1974), pp.241–73.
I. Lakatos, in J. Worrall and E. Zahar (eds), *Proofs and Refutations* (Cambridge University Press, Cambridge, 1976).
I. Lakatos and E. Zahar, 'Why did Copernicus' research program supersede Ptolemy's ?', in R.S. Westman (ed), *The Copernican Achievement* (University of California Press, Berkeley, 1975), pp.354–91.
L. Laudan, *Progress and its Problems* (Routledge & Kegan Paul, London, 1977).
D. Lecourt, *Marxism and Epistemology*, B. Brewster (trans), (New Left Books, London, 1975).
A.O. Lovejoy, *The Great Chain of Being* (Harvard University Press, London, 1964).
J.G. McEvoy, 'A "Revolutionary" Philosophy of science: Feyerabend and the degeneration of critical rationalism into sceptical fallibilism', *Philosophy of Science*, 42 (1975), 49–66.
E. McMullin, 'The history and philosophy of science: a taxonomy', in R.H. Steuwer (ed), *Minnesota Studies in the Philosophy of Science*, Vol.V (University of Minnesota Press, Minneapolis, 1970), pp.12–67.
P.K. Machamer, 'Feyerabend and Galileo: the interaction of theories, and the reinterpretation of experience', *Studies in the History and Philosophy of Science*, 4 (1973), 1–46.
B. Magee, *Popper* (Fontana, London, 1973).
H. Marcuse, *Eros and Civilization* (Sphere Books, London, 1969).
H. Marcuse, *An Essay on Liberation* (Penguin, Harmondsworth, 1972).
H. Marcuse, 'Herbert Marcuse on the need for an open Marxist mind', *The Listener*, 9 February 1978, pp.169–71.
P. Marris, *Loss and Change* (Routledge & Kegan Paul, London, 1974).
K. Marx, 'The eighteenth brumaire of Louis Bonaparte', in K. Marx and F. Engels, *Selected Works* (in one volume), (Lawrence & Wishart, London, 1970), pp.94–179.
J. Mepham, 'The theory of ideology in capital', *Radical Philosophy*, No. 2 (Summer 1972), 12–9.
J. Mepham, 'The structuralist sciences and philosophy', in D. Robey (ed), *Structuralism: An Introduction* (Clarendon Press, Oxford, 1973), pp.104–37.
A.I. Miller, 'On Lorentz's methodology', *British Journal for the Philosophy of Science*, 25 (1974), 29–45.
A. Musgrave, 'Logical versus historical theories of confirmation', *British Journal for the Philosophy of Science*, 25 (1974), 1–23.
A. Musgrave, 'The objectivism of Popper's epistemology', in P.A. Schilpp (ed), *The Philosophy of Karl Popper* (Open Court, La Salle, 1974), pp.560–96.
A. Oakley, *Sex, Gender and Society* (Temple-Smith, London, 1972).
T. Pateman, *Language, Truth and Politics* (Jean Stroud and Trevor Pateman, Sidmouth, 1975).
K.R. Popper, *The Logic of Scientific Discovery* (Hutchinson, London, 1959).

K.R. Popper, *The Poverty of Historicism* (Routledge and Kegan Paul, London, second edition, 1960).

K.R. Popper, *The Open Society and Its Enemies* (in two volumes), (Routledge and Kegan Paul, London, fifth edition, 1966).

K.R. Popper, *Objective Knowledge* (Clarendon Press, Oxford, 1972).

K.R. Popper, 'Indeterminism is not enough', *Encounter* (April 1973), 20–6.

K.R. Popper, *Conjectures and Refutations* (Routledge and Kegan Paul, London, fifth edition, 1974).

K.R. Popper, 'Replies to my critics', in P.A. Schilpp (ed), *The Philosophy of Karl Popper* (Open Court, La Salle, 1974).

K.R. Popper, 'The rationality of scientific revolutions', in R. Harré (ed), *Problems of Scientific Revolution* (Clarendon Press, Oxford, 1975), pp.72–101.

K.R. Popper, 'The logic of the social sciences', in T.W. Adorno et al., *The Positivist Dispute in German Sociology*, G. Adey and D. Frisby (trans), (Heinemann, London, 1976), pp.87–104.

K.R. Popper, 'Reason or revolution ?' in T.W. Adorno et al., *The Positivist Dispute in German Sociology*, G. Adey and D. Frisby (trans), (Heinemann, London, 1976), pp.288–300.

D.J. de S. Price, 'Contra-Copernicus: a critical re-estimation of the mathematical planetary theory of Ptolemy, Copernicus and Kepler', in M. Clagett (ed), *Critical Problems in the History of Science* (University of Wisconsin Press, Madison, 1959), pp.197–218.

H. Putnam, *Mind Language and Reality* (Cambridge University Press, Cambridge, 1975).

W.V.O. Quine, *From a Logical Point of View* (Harper & Row, New York, 1961).

J.H. Randall, 'The development of scientific method in the School of Padua', *Journal of the History of Ideas*, 1 (1940), 177–206.

G. Righini, 'New light on Galileo's lunar observations', in M.L. Righini Bonelli and W.R. Shea (eds), *Reason, Experiment, and Mysticism in the Scientific Revolution* (Science History Publications, New York, 1975), pp.59–76.

A. Ryle, 'The focus in brief interpretative psychotherapy: dilemmas, traps and snags as target problems', *British Journal of Psychiatry* (forthcoming).

K.F. Schaffner, 'Einstein versus Lorentz: research programmes and the logic of theory evaluation', *British Journal for the Philosophy of Science*, 25 (1974), 45–78.

C.B. Schmitt, 'Experience and experiment: a comparison of Zabarella's view with Galileo's in *De Motu*', *Studies in the Renaissance*, XVI (1969), 80–138.

D. Shapere, 'The structure of scientific revolutions', *Philosophical Review*, 73 (1964), 383–94.

D. Shapere, 'Meaning and scientific change', in R. Colodny (ed), *Mind and Cosmos: Explorations in the Philosophy of Science* (University of Pittsburgh Press, Pittsburgh, 1966), pp.41–85.

D. Shapere, *Galileo* (University of Chicago Press, Chicago, 1974).

W.R.J. Shea, 'Galileo's claim to fame: the proof that the Earth moves from the evidence of the tides', *British Journal for the History of Science* 5 (1970),

111–27.
W.R. Shea, *Galileo's Intellectual Revolution* (Macmillan, London, 1972).
Q. Skinner, 'Meaning and understanding in the history of ideas', *History and Theory*, 8 (1969), 3–53.
D. Stove, Review of K.R. Popper, 'The logic of scientific discovery', *Australasian Journal of Philosophy*, 38 (1960) 173–87.
E. Thompson, 'Romanticism, utopianism and moralism: the case of William Morris', *New Left Review*, No. 99 (Sept./Oct. 1976), 83–111.
S. Toulmin, *Foresight and Understanding* (Hutchinson, London, 1961).
S. Toulmin, *Human Understanding*, Vol. I (Clarendon Press, Oxford, 1972).
K. Tribe, 'On the production and structuring of scientific knowledges', *Economy and Society*, 2 (1973), 465–78.
H.V. White, 'Foucault decoded: notes from underground', *History and Theory*, 12 (1973), 23–54.
P.P. Wiener and A. Noland (eds), *Roots of Scientific Thought: A Critical Perspective* (Basic Books, New York, 1957).
R. Williams, 'Base and superstructure in Marxist cultural theory', *New Left Review*, No. 82 (Nov./Dec. 1973), 3–16.
R. Williams, *Marxism and Literature* (Oxford University Press, Oxford, 1977).
W.L. Wisan, 'The new science of motion: a study of Galileo's *De Motu locali*', *Archive for History of Exact Sciences*, 13 (1974), 103–306.
E.G. Zahar, 'Why did Einstein's programme supersede Lorentz's ?', Part I, *British Journal for the Philosophy of Science*, 24 (1973), 95–123. Part II, 24 (1973), 223–62.

INDEX

Absences, 32-4, 38
Abstract possibility, 193, 200, 207
Anti-psychologism, 90-4
Aristotelianism,
  absence of mathematics in, 33-4, 108, 186
  incommensurable with classical physics, 18
  motion a process, not a state, in, 184-6
  its structure, 162-8
  as structurally excluding inertial physics, 32-3
  shoring up its structure, 209

Bachelard, G., 17, 177
Bellarmine, Cardinal, 160
Bhaskar, R., 171
Bohm, D., 111
Bohr, N., 111-13
Braudel, F., 38, 208
Break, 15, 20
Buridan, J., 176-82
Butterfield, H., 16-20

Carnap, R., 117
Carr, E.H., 66
Cassirer, E., 33
Chalmers, A., 18
Clagett, M., 186-91
Clavelin, M., 31 *et seq*, 163-8, 177, 200
Colletti, L., 40
Contexts of Discovery and Justification, 12, 86, 92-4, 110, 121, 198
Continuity,
  Clagett and Shapere on, 186-93
  and development by accretion, 21, 34
  in Kuhn's 'normal' science, 22
  in Lakatos' methodology, 71
  Marcuse and, 147

  of medieval with classical physics, 9, 17, 175-8
  originality denied by, 34
  Popper's, 46-7, 64-6
  in revolutions, 18
  and subjectless epistemology, 192
  and tenacity, 71, 74
Copenhagen Interpretation, 111-13, 117, 140
Copernican Revolution,
  aesthetic considerations in, 158
  earth's movement central to, 157-62
  Feyerabend's account as one-sided, 121 *et seq*
  Lakatos and Zahar's account criticized, 94-100
  Toulmin on, 26-7
Cosmos, 33-5, 186, 204-6

Descartes, R., 183
de Solla Price, D., 157-8
Dijksterhuis, E.J., 18
Discontinuity,
  and 'archaeology', 32
  and changes in standards of explanation, 108
  and changes in ontology, 108
  and the contexts of discovery and justification, 199
  and disengagement, 128-9, 149
  Feyerabend and, 107-18
  gestalt switches as metaphor for, 24-6, 31
  and incommensurability, 10-11, 194-200
  Kuhn and, 21, 23
  of medieval with classical physics, 10, 16, 29, 175-8
  and originality, 17, 175-6
  Popper and, 45-7, 67
Duhem, P., 17, 58-60

INDEX

Drake, S., 17, 108, 177
Dray, W., 192
Dunn, J., 41, 191-2

Edgley, R., 41, 94, 173, 174
Einstein, A., 112

Feminism, 154
Feyerabend, P.K.,
  his conservatism, 142-5
  on the constitution of
    rationality, 136-8
  discontinuity and, 107-18, 128
  on facts as compatible with
    several theories, 112-18
  on incommensurability, 194-8
  his methodological anarchism
    and Dadaism, 118-21, 139,
    142-5
  on 'reason' as faith, 125
  on 'subjective' factors and
    revolution, 135-7, 145-8
  on the tower argument, 123-4,
    130-2
Foucault, M., 10, 18, 32, 65, 129,
  137

Galileo,
  claims originality, 175-6
  his cosmological imperatives,
    204-6
  not a Dadaist, 109
  and impetus theory, 203
  as methodological anarchist or
    Dadaist, 134-5
  on 'neutral' motion, 201-3
  and the principle of inertia, 35,
    178, 200
  rejects Aristotelianism as false,
    13, 149-52
  his relativity principle, 205
  and the telescope, 132-4
  his tidal theory, 206
  his tower argument, 130-2,
    169-71, 205
  and truth, 109, 161
Gravity inherent to bodies, 36,
  189, 207

Hall, A.R., 15-16, 26-8
Hanson, N.R., 24
Heisenberg, W., 66, 112-13
Hempel, C., 113
Hermeticism, 107
Hume, D., 50, 171

Ibsen, H., 36
Impetus, theory, 178 *et seq*
Impossibility, 13, 32-7, 152, 161-2
Incommensurability, 10-11, 18,
  194-200
Internal history, 88-93

Kant, I., 15, 19
Kepler, J., 57
Kierkegaard, S., 137
Koyré, A., 10, 17, 29 *et seq*, 163-8,
  177
Kuhn, T.S.,
  his discontinuous theory of
    history, 21
  gestalt switches in, 24-6, 31
  on incommensurability, 194-8
  'normal' science as continuous,
    22
  on revolution and paradigm
    destruction, 22-3
  revolutions and rationality, 25

Latatos, I.,
  his defence of Popper against
    Kuhn, 71-4
  on internal and external history,
    88-93
  and the liberal conception of
    revolution, 70-1
  and logic and psychology, 90-4
  and 'normal' science, 74-81
  and 'novel' facts, 82-4, 95-100
  and proliferation by fiat, 84-6
  on reason as slow and fallible,
    88, 100
  his spectator's perspective, 86
  and theory-elimination, 86-90
Laudan, L., 9
Lavoisier, A., 20
Lenin, V.I., 40
Limits, 35-7, 194, 209

Luther, M., 123

Mach, E., 117
Machamer, P.K., 149
Maier, A., 176
Marcuse, H., 147
Marris, P., 210
Marx, K., 14, 36, 147, 208
Mechanical philosophy, 107-8
Merleau-Ponty, M., 148
Momentum, 186
Mutation, 10, 17, 177

Nagel, E., 113
Natural necessity, 171
'Neutral' motion, 201-3
Newton, I., 57, 137, 182

Popper, K.R.,
    his anti-Utopianism, 65
    as conservative revolutionary, 48, 51, 66
    continuity and discontinuity in, 45-7, 64-7
    on the Duhem-Quine hypothesis, 58-60
    on 'holism', 58-61, 65
    his hostility to novelty, 56
    his 'third world' and tradition, 63
Principle of inertia, 9, 13, 30-5, 178, 182 *et seq*
Proliferation, 84-6, 116, 143

Quine, W.V.O., 58-60

Randall, J.H., 33-4, 64
Reason,
    and commitment, 13, 145-51, 219-20
    as constituted in different historical epochs, 136-8
    Dadaist view of, 109
    dialectical, 41, 94
    as differentiating, 120
    history as a laboratory for testing theories of, 91-3
    'liberal' conception of, 26, 28
    Popper's conception of rational social change, 65-6
    Popper's conservative concept of, 48, 61-7
    and reasoning, 25, 28, 40, 208-9
    works slowly and fallibly, 88, 100
Revolution,
    asymmetrical nature of, 12-13, 150
    a break with the past, 15, 19-20, 39
    as central to scientific advance, 9
    changes in meaning of, 19-20
    commitment during as combining conviction and openness, 145-51, 219-20
    commitment during as subjective, 135
    conceptual changes during, 9
    conservative aspect, in Popper, 48, 66-7
    continuing, 39
    and discontinuity, 9
    Einsteinian, 16, 31, 48, 53
    'liberal' conception of, 18 *et seq*
    'Marxist' conception of, 29 *et seq*
    a process, not an event, 11-12, 30, 37
    rationality of, 11-13, 40-1, 61-7, 151-2, 178
    rejection and replacement during, 11-12, 37
    in society and in science, 19, 28
    uneven development during, 13, 124, 135, 145, 218-19

Schmitt, C., 33-5
Schrodinger, E., 111
Specificity, 118-21
Shapere, D., 186-93

Tenacity, 74, 143, 149
Thompson, E.P., 147
Toulmin, S., 9, 26-8

Whig interpretation of history, 21, 30
Wittgenstein, L., 143-4

Zabarella, I., 33-4
Zahar, E.G., 94-100